AF708386

Dedication

Sam Folz
1923-1986

The National Corvette Restorers Society was founded in 1974 by seven enthusiasts frustrated by their inability to find accurate restoration information for their early Corvettes. Sam Folz was one of the original seven and served the NCRS as both technical director and president.

Sam's tremendous Corvette knowledge and, more importantly, his willingness to share it with all in his impeccably cordial manner, set the tone for the NCRS and its functions. Membership now exceeds seven thousand and those in the know acknowledge the NCRS as the premier organization of its kind and an instrumental element in the booming resurgence of interest in early Corvettes.

With respect and appreciation, *Corvette Driver/Owner Guide* is dedicated to Sam.

CORVETTE

Driver/Owner Guide
For 1953-1988 Models

MICHAEL ANTONICK

Motorbooks International
Publishers & Wholesalers Inc
Osceola, Wisconsin 54020, USA ®

First published in 1987 by Motorbooks International Publishers & Wholesalers Inc., P.O. Box 2, 729 Prospect Avenue, Osceola, WI 54020 USA

Printed and bound in the United States of America

Library of Congress Cataloging-in-Publication Data

Antonick, Michael.
Corvette driver/owner guide.

1. Corvette automobile. I. Title.
TL215.C6A56 1987 629.2'222 87-24690
ISBN 0-87938-273-2 (soft)

Cover photography by Michael Antonick

Motorbooks International books are also available at discounts in bulk quantity for industrial or sales-promotional use. For details write to Special Sales Manager at the Publisher's address

Credits

My thanks to all who contributed in any way to the information and photos contained in this book. Special thanks to:

John Amgwert
Alan Blay
Bob Bondurant
Kent Brooks
Dave Burroughs
Rich Chenet
Dave Friedman
Dick Guldstrand
Gary Lisk
Bill Locke
Bob McDorman
Doug Robinson

Auto Accessories of America
Chevrolet Motor Division
Eckler's Corvette Parts
Guldstrand Engineering
Lubrizol Corporation
Mid America Corvette Supplies
National Corvette Restorers Society
National Council of Corvette Clubs

Corvette! America's Star Spangled Sports Car, Karl Ludvigsen/Automobile Quarterly
Corvette! The Sensuous American, Michael Antonick/Michael Bruce Associates, Inc.
Illustrated Corvette Buyers' Guide, Michael Antonick/Motorbooks International
The Corvette Black Book, Michael Antonick/Michael Bruce Associates, Inc.

Mike Antonick

Contents

Introduction

"I want it."

If Corvette folklore is accurate, these were the words spoken by Ed Cole when he first viewed a secret General Motors sports car prototype in 1952. The car was a one-off, two-seater intended for display in the 1953 edition of the Motorama, the extravagant traveling car show GM moved around the country from 1949 through 1961.

The "I" in Mr. Cole's statement was used in the corporate sense. He was speaking for the Chevrolet Motor Division of which he was then general manager. He "wanted" this secret car, later to be named the Corvette, to be produced in volume for sale to the public. And he wanted a Chevrolet nameplate on it.

He got it.

Cole's underlying motives cannot be verified, for no one thought to ask the retired General Motors president before his 1977 death at the controls of a private airplane near Mendon, Michigan. But Chevrolet's initial actions spoke clearly. Chevrolet *selected* the individuals it wished to own Corvettes during the first production year, 1953, and it selected people of status. The Corvette was thus an image car, a magnet for drawing customers to Chevrolet showrooms. For all the years since, competing dealers have longed for their own Corvette-like attraction. It's been tried—from Studebaker's Avanti to Ford's Pantera to Cadillac's Allante—but the Corvette remains the image marque the others aim for.

For years, it's been said Chevrolet lost money with every Corvette it sold. In the early days, maybe so. Retired Chevrolet employees who've been asked about it say they suspect the Corvette *was* a profit drain. But there was so much corporate profit, and bookkeeping techniques for allotting profit and loss to specific models were so imprecise, no one knew. Or cared. The Corvette played its image role so well, it simply didn't matter.

Image isn't enough in today's competitive market. The Corvette enterprise is now a profitable one. If it ceases to be for long, the Corvette will cease to exist. That's one reason the current production Corvette is so good . . . it has to be in order to compete, to sell and to generate an acceptable return on Chevrolet's investment.

The image Chevrolet nurtured for the Corvette was one of performance. In the process, the Corvette developed a nearly mystical appeal.

A personal story: I was a teenager when the Corvette Sting Ray was unveiled in the fall of 1962. For the next few years, my car-loving friends and I spent countless hours at the local Chevy establishment drooling over every new Corvette arrival, each of us dreaming his own personal fantasies. One evening in 1963, the transport driver who'd just unloaded a Corvair for our local dealer, told us the big-city dealer forty miles away had three new Sting Rays on display. We pooled our pocket change for gas and drove over.

We arrived near closing time, and some of the showroom lights were already off. While my friends told the salesman our rich dads were willing to buy each of us a new Corvette provided there was an adequate group discount, I climbed into a black-on-black 1964 coupe parked in the far corner of the showroom. With the Corvette's tinted windows up, in the dimly-lit showroom, I was in another world.

Over two decades have since passed, but I clearly remember thinking if only I could have this car, all my problems would vanish. Any fair maiden I desired would be mine. She and I and it would be happy for eternity.

Life should be this simple.

Three years later, I had my Sting Ray. It did bring me great joy, and I look back fondly on our times together. It did not, however, solve all my problems. In fact, it brought unexpected new ones that someone forgot to mention in the resplendent sales brochures.

The expectations of Corvette ownership are so grandiose, especially for the young, that disappointment is nearly assured. Judging from past owners, though, we'd never guess it. You've heard the testimony:

"I wish I'd never sold the Corvette."

"I'd give anything to have it back."

"I'm looking for one exactly like the one I had."

Hindsight is so forgiving. Fact is, these people did get rid of their Corvettes, for reasons that seemed pretty solid at the time. Sure, their needs may have changed. They might have been drafted into the service. They may also have gotten fed up with water leaks, a harsh ride ("*she* made me sell it"), outrageously expensive insurance, back-breaking seats, heavy steering (no need for Nautilus tonight), a cramped cockpit, high maintenance cost, high fuel cost, poor ventilation, illegible instruments, fogging windows, a buzzing shifter, frame rot, soft-top-flop, vibration, wind noise, brakes that broke, or lying in the snow or mud to retrieve a spare tire.

Corvettes have never been perfect. Of course they can't live up to the preconceived expectations of many owners. Anyone can own and drive a Corvette. That's the easy part. Driving a Corvette to its and your potential, thoroughly understanding and enjoying Corvette ownership . . . that takes some effort.

And some help. *Corvette Driver/Owner Guide* was written to enhance the Corvette driving and ownership experience. Before I explain what this book is, let me tell you what it is not. It is not an encyclopedia of facts. It is not an investment guide. Nor is it aimed at Corvette owners who intend to store their prize on blocks waiting for it to appreciate.

There's nothing wrong with storing a Corvette away, and certainly nothing wrong with looking at a Corvette's appreciation potential before deciding to purchase. Storage techniques are discussed in this text because owners who drive their Corvettes might still need to store them at times. And Corvette models with the best driving attributes are often those that appreciate the fastest.

But *Corvette Driver/Owner Guide* is primarily for owners who want to drive their Corvettes. Yours may be a show-quality Corvette that sees the street only on sunny weekends, a restored vintage racer, a daily driver new or old, or anything in between. You may wish to cruise the neighborhood at 25 mph, or unleash your machine to its limits. Corvette drivers, this book was written for you.

It's not necessary for you to already own a Corvette to find this book useful. If your intention is to drive the Corvette you buy, this book will be very helpful in aiming you toward the model with the driving and owning characteristics that best fulfill your expectations.

You see, the Corvette is many different cars. Granted, each has been a two-seater with a plastic body. Each has had an engine in front, driving the rear wheels.

Next, you expect to hear that the difference between a 1953 and one of today's models is like day and night. True enough. Surely it's no surprise that the four generations of Corvettes are vastly different automobiles. The 1953-62 Corvettes all had solid rear axles and personalities much different from the independent-rear-suspension Sting Rays introduced in 1963. The 1968-82 models shared major suspension components with the 1963-67 Corvettes, but had completely different bodies and interiors. Then came the 1984 and newer Corvettes, and they're very unlike anything before.

Here comes the difficult part. Corvettes differ significantly *within* the same model year, let alone the same generation. Of the thirty or so Corvettes I've owned, my all-time favorite is a 1967 I still have. My least favorite was another '67 I disposed of in a week.

Permit me to clarify. Sting Ray lovers don't believe Chevrolet made a bad Corvette in 1967. Some owners, intending to make them better, ruined them. So it was with the '67 just mentioned. But this brings up an important point. All Corvettes have shortcomings. How they're corrected, if corrected at all, can greatly affect a Corvette's value, both positively and negatively. *Corvette Driver/Owner Guide* isn't an investment guide, but it won't ignore the effects of modifications on value. For instance, stainless steel inserts added to the brake calipers of 1963-82 Corvettes are a sure fix for a common corrosion leak problem. This repair enhances a Corvette's resale value and is not considered detrimental to originality by any judging organization. On the other hand, changing some Corvette handling traits requires suspension modifications. Alterations like these affect a Corvette's originality and may detract from its value. The impact on value of modifications will be explained so that you can make sound decisions for your own needs.

Regardless of your intentions, you should know the importance of originality to Corvette enthusiasts. To the Corvette elite, originality isn't merely having correct motor numbers and exterior paint colors. It goes well beyond that. For an unrestored Corvette, it means maintaining the car in the same state as when it left the factory gates. For restoration, it means putting the Corvette back to the factory-gate condition. Sounds simple enough, but it's not. Serious Corvette restorers reproduce the stenciled numbers on frames, they overspray black-out paint on certain components, they don't smooth the waves typical of fiberglass panels, they duplicate the paint run patterns on parts the factory originally dipped. You might think this is nuts and not for you, but you should keep tuned into the hobby well enough to make intelligent decisions when it comes to your own Corvette. Some modifications are double losers. You lose what the modification cost, and the Corvette's value is decreased by what it would cost to undo the modification.

Why does the Corvette community place such emphasis on originality? Part of it is backlash to an era when everyone "personalized" their cars, especially Corvettes, in ridiculous and tasteless ways. Also, Corvettes have always been favorites of thieves and strippers, and no one likes the idea of owning a Frankensteinian Corvette that's been stitched together from three or four carcasses.

But there is a more fundamental reason. Factory-built originality is a nonmoving target, not subject to whims or trends. They built it the way they built it, period. The arguments about little nuances of factory techniques go on forever—it keeps the Corvette hobby alive and healthy—though the range of debate is reduced continually as Corvette restorers ferret out finer and finer details.

It's possible to make changes without killing a Corvette's value. Most of the modifications pointed out in this book are reversible, so that you, or a future owner, could take the Corvette back to a factory-original condition. For instance, two Corvettes, one a 1964, the other a 1967, recently sold for extraordinary prices even though their owners had gutted each for racing shortly after purchasing them new. Both owners had the foresight to save the original components. The 1964's leather interior had never been used. The 1967's big-block engine had been crated and had only a few hundred miles of use after two decades!

Much of this book's text is devoted to Corvette shortcomings and problems, so it's inevitable some readers will perceive a negative tone. That bothers me. Corvettes are magnificent automobiles. Yet, because they accomplish their performance mission so well, drivers have always had to pay some price in comfort and convenience. Other concerns, like insurance, theft and fuel requirements, aren't the car's fault, but are real-world problems many owners have to deal with.

We also can't forget that some Corvettes were built in entirely different eras. Take seat design. At National Corvette Restorers Society meets, an award is presented to the owner who drives a vintage Corvette the longest distance to get there. When it's announced that so and so drove 1,800 miles in his 1955 Corvette, there's always a collective gasp from the audience. Can the guy still walk?

But considering the time frame in which these cars were built, they're not so bad. Listen to what *Road & Track* magazine, often quite critical of domestic automobiles, said in its July 1955 issue about the 1955 Corvette: "Ride qualities are excellent and directional stability at high speeds is near-perfect. The seats are beautifully done and very comfortable . . ."

Dick Guldstrand, the Corvette suspension genius who is quoted extensively in Chapter Three, now "cringes" to think he actually raced the unsophisticated 1957 chassis. In its day, nothing could touch it.

As they say, it's all relative. The first jet was impressive in its day, too.

The key to enjoying the Corvette driving and ownership experience is knowledge. The Corvette isn't a mysterious car, but it is different. The problems associated with Corvette ownership aren't insurmountable, but they're real and have to be understood to be overcome.

Does it take extra effort to master Corvette driving skills and ownership requirements? Sure. Is it worth it? Absolutely.

The Corvette is special. It has been right from the start. Ed Cole sensed it. But over the years, the Corvette has evolved into so much more than a great high performance sports car. It is the ultimate cult car. Not even Ed Cole, who wanted the Corvette for his Chevrolet Division, could have imagined how universally *wanted* the little two-seater would become.

1
Ergonomics

Ergonomics is one of the popular buzz words that has cropped up in auto jargon in recent years. Let's find out what it means and then see how well different Corvette models measure up.

The noun "ergonomics" is derived from *ergon*, the Greek word for work. *Webster's New Collegiate Dictionary* says ergonomics means biotechnology, and defines biotechnology as "the aspect of technology concerned with the application of biological and engineering data of problems relating to man and the machine."

In the auto industry, ergonomics refers to the design of an automobile's interior in a way that best permits the driver to accomplish the "work" at hand, driving. Many things are involved, including how the seats position the driver, the instrumentation, the controls, adjustments and visibility. This all sounds straightforward enough and, in some respects, it is. There are certainly automobiles built today with superb driving environments.

A car doesn't have to be new to be good, or old to be bad. Some old cars are quite good. And, in spite of today's emphasis on ergonomics, many manufacturers, especially domestic, continue to design vehicles that are ergonomic disasters. Being kind, one could say the problem is the oil and water blending of two very different sciences, biology for the human aspects of the driver, and engineering for the nuts and bolts of the hardware. Being honest, some chauffeur-driven executive overrules good design because he *knows* gimcrackery is more important.

All things considered, Corvettes stack up pretty well in the ergonomics department, especially considering each model's build era. Corvettes were "sporty" right from the start, and performers almost from the start, and these attributes dictated consideration for the driver's task. But there were glaring deficiencies designed into many Corvettes. Ergonomics is one area where an owner's ability to make corrections is limited, but some fixes are available.

Seating

It's said the most sophisticated chassis is useless without proper tires, because they're the car's only contact with the road. The driver's main contact with the car is the seating, and few things will influence a driver's ability to drive properly as much as the seat.

Seats are normally thought of as being comfortable or not, but ergonomics and comfort aren't the same. Ergonomics refers to the man-machine interface. The seat in an Indy racer is ergonomically superb because it holds

Top right: The seats in 1953-55 Corvettes were well made but positioned the driver's legs very close to the steering wheel. Seat adjustments were limited to fore-aft movement for the driver seat only.
Left and bottom right: Both driver and passenger seats in 1956-62 models (1961 shown) had fore-aft adjustments. Aesthetically, all 1953-62 Corvette seatbacks blended beautifully with convertible top lids, so aftermarket seats look out of place.

---

the driver in the best position to accomplish the driving tasks, considering the constraints imposed by the vehicle's configuration. A Cadillac's bench seat would be more comfortable, but it might get low ergonomic marks.

Today, auto companies intelligent enough to pay attention to ergonomic seat design inevitably come up with comfortable seats as well. It's the reverse that often isn't true. For now, know that this chapter deals with ergonomics, the next with comfort. The seats of some Corvette models had poor ergonomic designs, but were still comfortable.

When the first Corvette arrived, ergonomic design considerations were in the dark ages. In 1953-55 models, the only provisions for adjustment were for fore and aft seat travel, and only for the driver seat. Starting in 1956, driver and passenger Corvette seats had fore and aft adjustments.

Left: Though minimal, 1965-66 seats (1965 leather shown) did have side bolsters to restrain lateral movement.
Top center: Lateral support was reduced in the flatter 1967 seat, but the seat's construction was very durable. Headrests were optional.
Bottom center: Seats in 1968 model could cause "submarining" toward firewall because they were slippery and had increased back rake. Headrests were still optional in 1968, but were designed into the seats starting in 1969.
Right: Unusual seat design introduced in the 1978 Pace Car (1979 shown) offered excellent lateral support for occupants narrow enough to fit.

All 1953-62 Corvette seats were vinyl-covered with flatish seating surfaces. The combination of vinyl and flatness resulted in much slipping around in hard cornering. Paul Reinhart, who raced a production 1957 Corvette to regional Sports Car Club of America (SCCA) California championships in 1960, 1961 and 1962, remembers the seat in his 1957 as a real problem because of its slipperiness. Stock class rules required production seats, so as Paul put it, "I pulled the belt as tight as I could stand, held on to the wheel, and lived with it."

Though all the "classic" Corvettes shared similar chassis construction, the seats did vary. Think of the 1953-55 models as one group, the 1956-62 models as another. One sits *on* all these seats rather than *in* them, but the seats of the 1953-55 Corvettes were better because they were softer and wider, resulting in some side bolstering effect.

Why not put the earlier seats in the later models, you ask? They won't fit. The seats were made narrower in 1956. This model got roll-down windows, definitely a needed feature, but the window forced the door section to get wider, eating into seat width. Also, a heavier top bow structure in the 1956-62 models intruded on seat space.

Putting a non-Corvette seat in the 1953-62 models doesn't work well either. These Corvettes were all roadsters and they look as if they were designed with the soft tops folded away. The top surfaces of the seatbacks are sculptured to blend beautifully with the convertible top lid. Changing to any other seat ruins the aesthetics. If it's a racer, aesthetics may not matter.

For street use, you'd be displeased with the appearance of anything but factory seats.

If the slippery nature of these seats is particularly bothersome, or if perspiration glues you to the vinyl surface, consider sheepskin slipcovers. They'll protect the original seats and help, but not completely cure, the sliding problem. The driver won't slip on the cover, but the cover will move on the seat somewhat. Most dealers don't stock these low-demand covers, but there are manufacturers who will gladly custom-make sheepskins for virtually any automobile.

Another modification an owner can make to the 1953-62 seats is to lift or rock a seat by shimming between the floor and the seat's mounting brackets. Most drivers find the steering wheel is too close, especially in 1953-57 models, so shimming the front of the seat to rock it back unwinds the elbows, and creates better thigh support. The Catch-22 is that the steering wheel diameter was so large, some drivers have trouble getting their legs between the wheel and seat surfaces to start with. Shimming the forward part of the seat aggravates this further.

Ergonomically, the seats in 1963-67 Corvettes were quite good. They were available in vinyl or leather. Both surfaces were still slippery, but these seats were closer to true "buckets." The 1965-66 seats were the best in this regard, the unique '67 seat nearly as good, the 1963-64 less so. One advantage of the 1963 seat was a mounting bracket design that permitted seat cushion adjustment without the shimming described for earlier models. Problem was, the adjustment required wrenches, a "take it to your dealer" feature. Any adept owner could do it, but few knew about it. Instead of modifying the brackets to be more easily adjusted with a knob or lever, then promote what was an excellent idea a decade ahead of the industry, Chevrolet just deleted it from Corvettes the following year.

The 1963-67 era saw the introduction of head restraints. They were options in both 1966 and 1967 Corvettes, but scarce at less than four percent installation the first year, less than eight percent the second. Head restraints were eventually mandated by the federal government for their ability to minimize whiplash neck injuries in rear collisions but, with proper design, they serve an ergonomic function by supporting the driver's head in a proper position and preventing neck and shoulder fatigue. This is more important as the seatback angle increases and thus is not a big factor in 1963-67 models since their seatbacks are quite upright.

These seats do not blend into the convertible top panels like earlier Corvettes and, of all Corvette models, best lend themselves to aftermarket seat replacements. Companies like Recaro offer off-the-shelf brackets and an assortment of orthopedically superb seats with multiple adjustments that will bolt right into any 1963-67 Corvette. They wouldn't be right for a concours stressing factory originality, and factory seats are plenty good enough for the casual use most owners give these models. But a high-quality aftermarket seat is the only way to go for someone who wants to do either spirited or long-distance driving. These midyear Corvettes are enthusiast dream cars, because they were so far ahead of their time. Except for a couple of items, seats being one, these cars compare well to anything on the road today.

Next comes 1968-78 seating (except for the 1978 Pace Car replica which belongs with the 1979-82 Corvettes). Some enthusiasts, this author

Left: Here are Recaro seats in a 1964 Corvette. They look great and have multiple adjustments, but this particular seat's width limits fore-aft travel because of interference with the transmission tunnel.
Right: Recaro seat installation in 1973 model creates a superb driving environment. Beware of reduced headroom, however. Recaro photo.

---

included, consider these the poorest Corvette seats ever. The chassis was the same as 1963-67, but the driving environment suffered because of the 1968's body and floor panel redesign. For starters, the optional three-speed Turbo Hydra-matic transmission was bigger than the two-speed Powerglide it replaced, and required a wider tunnel and narrower seats. Since the 1968's roofline was lower, seatbacks were raked an additional five degrees (thirty-three degrees compared to twenty-eight degrees in 1967) to gain headroom.

The more a seatback is raked, the more critical the seat design becomes because it must prevent the driver from sliding, not only from side to side, but also from "submarining" toward the firewall. The 1968-78 seats didn't do well. Anyone who's spent considerable time in one of these models knows it's a battle to keep propped up in the seat. Vinyl was finally banished from Corvette seating surfaces starting in 1977 when the choices became cloth or leather. Cloth seats of 1977 and 1978 were less slippery than leather or vinyl, but cloth only partially cured the problem.

Chevrolet rolled out its Pace Car replica Corvette in mid-1978 and it included a new, significantly improved seat. This Pace Car seat design became standard in all Corvettes starting in 1979 and continuing through 1982. It represented Chevrolet's first real ergonomic thought in Corvette seat design. Big, thick side bolsters restrained side-to-side occupant movement, though the combination of a narrow seat overall and a thick bolster made for a small butt-pocket. This seat either fit, or it didn't. It also had a peculiar back that hinged about a third up from the base instead of at the base. The break corresponded to the rear compartment ledge so that you didn't need to fold as much seat forward to get at the stowage area, and what you did, folded out of the way without hitting the instrument panel. Unfortunately, the break also corresponded to a sensitive portion of some drivers' backs. This seatback design ruled out an effective rake adjustment, so the Corvette sailed right through 1982 without ever having had a properly adjustable seatback.

Nineteen eighty-one did see one nice seat option, a power unit for the

The leather surfaces often didn't wear well (cloth versions were more durable but not as attractive), but the seats of 1984-88 Corvettes were ergonomically superb. A sport version had controls on the side bolsters for bolster position, back rake and pneumatic lower-back adjustment.

---

driver side. It was GM's six-way power base which moved the entire seat fore and aft, full up and down, or front and rear up and down (tilt) independently.

Some remedies: The owner of a 1978 Pace Car and 1979-82 model is probably content with the stock seat, but there is the option of an aftermarket product. The choice is more limited than with the 1963-67 series because of the narrow space available. Aftermarket seats nearly always require adapting brackets, so headroom can be reduced. If an aftermarket seat is tempting, work out a trial installation before purchase, because of the potential fit and headroom problems. In order to get the necessary headroom, some owners have had to bolt aftermarket seats in without the fore and aft slide rails. That may be too much of a compromise.

The owner of a 1968-78 (non Pace Car) has two options. There's the aftermarket seat, or the better factory seat from a 1979-82 Corvette. The latter means locating a used set at a salvage yard, Corvette show, or in ads. They're fairly plentiful and should cost no more than half that of good aftermarket seats, though they'll no doubt have to be re-covered.

The owner of a 1984 or newer Corvette doesn't have to go looking for seats. Not that what he has is perfect. The leather and cloth materials didn't

 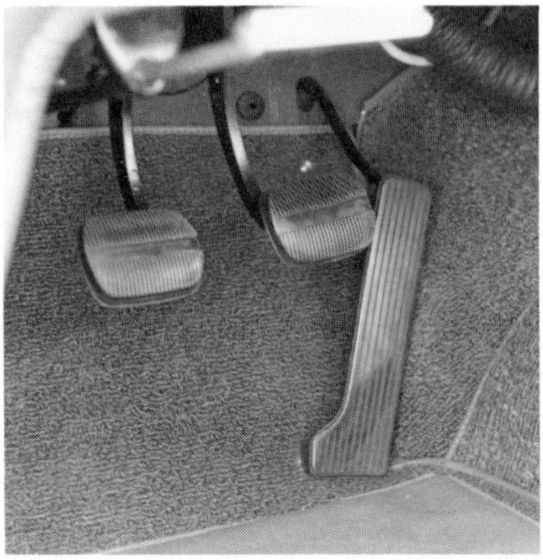

Left: They may be old cars, but the pedal locations in early Corvettes were terrific. This is a 1955, the only year of the 1953-55 period when a manual shift was available.
Right: Wide footwells and perfect spacing made the pedal setups in 1956-62 models great for heel-and-toe downshifting.

---

wear well, and the foam was too soft. But the seat design was otherwise ergonomically excellent. In the 1984-88 period, two versions were available, a standard and a sport. Depending on the year, each was available in cloth or leather, and power base units were optional for either type (power bases for passenger seats weren't available until 1987).

These seats were characterized by high, thin side bolsters, adjustable seatbacks, correct contouring, even wool pads under the cover surfaces for moisture dissipation. The seatback adjustment was manual for the standard seat, electric for the sport. The sport also featured three pneumatically operated lumbar cushions for fine-tuning lower back support and electrically-operated bolsters to accommodate different driver widths. Of all the sport seat features, the electric rake adjustment was most useful because it was so convenient, and nothing relieves long trip fatigue better than a slight rake angle change. The manual worked almost as well with minimally more effort.

The air cushions were a nice idea but weren't as effective as Recaro's aftermarket units. The electric bolster adjustment was a waste of hardware; a manual adjustment would have been fine for something changed so infrequently.

Pedals

Another important ergonomic consideration is how well the driver can operate the pedals, especially in manual-shift models. Two important design considerations are the distance between pedals and the depth differential between the brake and accelerator. The distance between pedals shouldn't be so close that two pedals are unintentionally pressed at once or that one pedal is mistaken for another, nor so far that heel-and-toe

Top left: Footwells were made narrower for 1963, but pedal positions were still good, especially with power brakes because the brake pedal was closer to the floor for easy heel-and-toe downshifting.
Top right and bottom left: Footwells were narrower again for 1968 as these 1971 (bright pedal trim) and 1979 model photos show.
Bottom right: Pedal spacing for 1984-88 models was good, given the limited space available. These were the first Corvettes with a dead pedal for the clutch foot.

maneuvers aren't easily accomplished. The depth differential between the brake and accelerator is important for heel-and-toe operation. The two factors are interrelated because, generally speaking, the closer the pedal spacing, the closer the brake and accelerator depth differential has to be for good heel-and-toe technique.

Another element of good pedal design is a dead pedal, a resting place to keep the driver's clutch foot off the floor and closer to the clutch pedal. The best dead pedals are rigid, covered with hard, slip-resistant rubber like the functional pedals. At first thought, it might seem a dead pedal is only useful for manual-shift cars. It's not true, because a good dead pedal also serves as a sort of brace point for the left leg. It can be useful in hard cornering or to help readjust the driver in the seat.

The functional pedal arrangement in Corvettes has always been good, the dead pedal provisions either nonexistent or fair. Corvettes built through the 1982 models had no dead pedals, though the floor panels of the 1968-82 models did have an elevation which would accommodate the clutch foot. Unfortunately, in 1968-76 Corvettes, that's where the headlight dimmer

Top left: Steering columns for 1953-57 models (1957 shown) didn't adjust and the column length put the wheel uncomfortably close to some drivers.
Bottom left: Steering column for 1958-62 models (1958 shown) didn't adjust either, but Chevrolet shortened the column three inches in 1958 to put the wheel closer to the instrument cluster.
Right: The telescopic column adjustment, shown at each extreme in these 1965 model photos, became optional in 1965. Steering column length could also be changed with an under-hood adjustment for all 1963-66 models.

---

switch was, hidden under the carpeting. Starting with the 1984, Corvettes did have a dead pedal, a ledge in the floor pan for the left foot to rest. It wasn't as effective as it could have been, because its surface was a vinyl pad formed into the carpeting rather than a rigid block.

Owners of earlier Corvettes who've felt the need for a dead pedal have simply fashioned one from a block of wood and a pedal pad.

No Corvette models have functional pedal arrangements so poor that modifications are necessary, but some are definitely better than others. The first Corvette to offer a manual transmission was the 1955 model. These and all Corvettes through 1962 have relatively wide footwells and excellent pedal locations. Power brakes were never available on these models, so the manual brake pedal is considerably higher than the accelerator, but the wide footwell allowed adequate pedal spacing that was fine for heel-and-toe operation. Hundreds of these Corvettes have raced with no pedal modifications whatever.

Top: A tilt-telescopic column became a Corvette option in 1969 (1972 and 1977 versions shown).
Bottom: These photos display the degrees of adjustment possible with the superb tilt-telescopic column of 1984-88 models (1985 shown).

The pedal placements in 1963-82 Corvettes were also very good, but there was less room to operate in, due to a footwell that was narrowed in 1963 and again in 1968. Power brakes became available in Corvettes for the first time in 1963. Since the power brake pedal was closer to the firewall, any of the 1963-82 models with power brakes is a breeze to heel-and-toe (power brakes were standard in Corvettes starting with the 1977 model). The manual brake models are more difficult to heel-and-toe and can present difficulty for some owners. One solution is to bend the brake hanger rod, but an easier and less permanent one is to fashion a removable spacer block for the accelerator, provided the reduced legroom is acceptable.

Steering Wheel Adjustability

The adjustability of the steering wheel is another important ergonomic concern. Here, Corvette models run the full gamut from a zero rating to a perfect score.

Steering wheels in 1953-62 models have no adjustment provisions. Drivers complain most about the 1953-57 Corvettes because the combination of a too-long steering column and the wheel's seventeen-inch diameter puts the steering wheel practically in the driver's lap. Many who raced the

1956 and 1957 Corvettes shortened the steering column three inches to get better arm extension. Chevrolet recognized the problem and shortened the column, starting in 1958. This helped, but there still were no adjustments.

The steering wheels of 1963 and 1964 Corvettes didn't have driver adjustments, but the fore and aft wheel position of 1963-66 models could be changed with an under-the-hood steering shaft adjustment (another take-it-to-your-dealer feature). The columns in all the 1963-67 models entered the cockpit at an angle, and this might irritate drivers initially, but the steering wheel location was good (some auto writers thought it too high), and clearance between the wheel and seat was generally adequate. As with the earlier Corvettes, most drivers can adapt.

The first real adjustable steering wheel for Corvettes came along as an option in 1965. It was RPO-N36, and would telescope the steering wheel over a range of three inches. This option could be ordered in all 1965-68 models. Even though this option was strictly telescopic, not tilt, enthusiasts value it highly. Changing a nontelescopic model over to telescopic requires changing the steering column and hub assembly, so this means locating a used one at a salvage yard (unlikely), at a Corvette show, or in ads. Be prepared to pay at least $600.

Nineteen sixty-nine brought an entirely new column for Corvettes, the locking type which incorporated the ignition switch. This new column brought with it a tilt-telescopic option. If only the earlier models could have had this! From 1969 on, steering wheels in Corvettes equipped with the tilt-telescopic feature were ergonomically excellent. The column with tilt-telescoping adjustment was optional (RPO-N37) from 1969 through 1979, then became standard Corvette equipment in 1980.

Instrumentation

Most enthusiasts rank instrumentation near the top of their ergonomic considerations. They'd be right, too. Drivers hear and feel what their automobile is doing to a degree, but instruments provide the definitive telemetry.

Good instrumentation involves at least three basic criteria. First, the instruments must convey all the commonly accepted important automotive functions and do it in a measured way. Second, instruments must be located so they're easily visible. Third, instruments must present their data in a way the brain can interpret quickly.

Many American autos have only a speedometer and fuel gauge, relying on warning (idiot) lights for monitoring engine functions. It's cheaper, of course, but manufacturers justify it by citing statistics to prove there are less catastrophic engine failures in cars with warning lights. With an event such as a sudden loss of oil pressure, it's a valid argument. The problem with warning lights is their inability to signal a problem as it's developing. Warning lights in combination with full instrumentation is the obvious solution, but too few manufacturers have opted for it.

The criteria for proper instrumentation are no secret, yet few automobiles have genuinely good instrumentation. Some Corvettes are among the best, but not all shine.

To Chevrolet's credit, all Corvettes have had complete instrumentation. Even the very first Corvette model, with its six-cylinder engine and two-speed automatic transmission, had a tachometer in addition to speed-

Bottom: Except for 1957's RPO-579E, all 1953-57 models shared this instrument layout with a central-dash location for the tachometer.
Top left: Here's the tachometer setup with the factory racing option, RPO-579-E.
Top right and center: These are non-stock owner solutions to the instrument shortcomings of early Corvette models.

Top left: The tachometer and secondary engine instruments moved front and center in 1958's redesign (1961 shown).
Top right: The new 1963 instrument placements were excellent. The only valid criticism of the instruments was the too-fancy face designs.
Bottom: Instruments of 1965-67 (1966 shown) left nothing to criticize.

---

ometer, oil pressure, ammeter, coolant temperature and fuel gauges, and a clock. Very aircraft-like, the tachometer even displayed the cumulative total engine revolutions, a neat feature that continued through the 1957 model.

The deficiencies occurred in where the instruments were placed, or how they presented data.

In 1953-57 Corvettes, the tachometer is positioned in the center of the dash, out of the driver's normal line of sight. Close to it, nearly as difficult to see, were all other instruments except the speedometer. The speedometer was where it belonged, right in front of the driver. Chevrolet knew it had a problem when it got racing fever in 1956, but magazine writers spotted it earlier. In its July 1955 review of the 1955 Corvette V-8, *Road & Track* said, "The instrument panel of the Corvette commits a cardinal sin by using a very small tachometer and placing it in the center of the instrument panel rather than alongside the speedometer."

While designers quickly worked up a new instrument panel for the 1958 Corvette, a stop-gap solution was introduced for 1957 as part of option RPO-579E. It was a big tach that crudely bolted right on top of the steering column. A round Corvette emblem covered the tach hole left in the center of the dash. Many, many owners of pre-1958 Corvettes have added

Left: Secondary instruments moved back to the center-dash area in 1968 (1970 shown) with just the tachometer and speedometer staying in front of the driver. Right: The center console was redesigned for 1977; the speedometer and tachometer area for 1978. This 1982 Collector Edition model photo shows both new designs.

---

tachs in the same location. It's incorrect for concours, of course, unless you're fortunate enough to own a real RPO-579E 1957.

The new instrument cluster arrived in 1958 Corvettes and lasted through 1962. The tach was too small, its needle bounced too much, and the cluster pod's design was clunky, but all the instruments (except the clock which was in the center console) were right in front of the driver where they belonged. Remember that Chevrolet shortened the steering column in 1958. This, combined with the steering wheel's seventeen-inch diameter, gave the driver a completely unobstructed view of the speedometer and tachometer. The secondary instruments could be blocked by the steering wheel's ten o'clock and two o'clock spokes, however. Instrument faces were redone in 1960 for better legibility.

For top-notch ergonomic instrumentation, check any 1963-67 Corvette. Writers have speculated that this instrument panel was designed by people who had no idea what the exterior designers were doing, because the dash's twin pods bore no relationship to any exterior styling theme. Maybe so, but once inside, the instrument presentation was hard to beat. The big speedometer, big tachometer and all secondary engine instruments were placed in near-perfect viewing positions. The instrument faces in 1963, and to a lesser degree in 1964, were too gaudy, but this didn't much affect their legibility. In the 1965-67 models, the instrument appearance and function were just plain magnificent. Few sports cars measure up to this standard.

The body for the 1968 models included a new interior and instrument cluster. The big-tach-and-speedometer theme was retained—they stayed right in front of the driver. But the secondary instruments all moved to the center console area. The good side of this was that the secondary instruments were never blocked from view by a steering wheel spoke. But ergonomically, a driver shouldn't have to turn his head, only drop his vision momentarily, to view any engine- or performance-monitoring instrument. The designers responsible for the central instrumentation locations in the 1953-57 Corvettes must have retired, because here it was again. The arrangement introduced in 1968 continued through the 1982 model, though

It was criticized for its electronic speedometer and tachometer displays, but the 1984-88 instruments did present a comprehensive array of information. Secondary instruments were backed up by warning lights, and all instrument locations were excellent. Chevrolet photo.

1977 models received redesigned and more legible secondary instruments. A similarly redone tach and speedometer followed a year later.

Auto journalists couldn't find much in the ergonomic aspects of the 1984 Corvette to criticize, but the "video game" instrumentation was the exception. It did get high marks for the amount of information presented. In addition to all the data previous Corvettes communicated to the driver, the 1984 Corvette added oil temperature, instantaneous fuel mileage, average fuel mileage and the estimated driving range before refueling. The secondary instruments' digital readouts were backed up by warning lights. Finally!

The locations were good. The tachometer and speedometer were in the main instrument cluster; other displays were centered in a pod in the upper portion of the cluster.

The shortcoming was in how the information was delivered to the driver, and the driver's resulting difficulty in reading the instruments at a glance. The main culprits were the speedometer and tachometer—especially the tachometer. The speedometer and tachometer displays each had a graphic band (the engine's torque curve in the case of the tach) which illuminated as the car's speed or engine rpm increased. Each also had a digital readout, up to three digits for speed, two for the tach. Drivers ignored the graphic speed band, preferring the digital readout. With the tach, just the opposite. Two digits didn't relate well to thousands of rpm, so the driver was left with the graphic band. It was difficult to interpret quickly and accurately.

Control Locations

An area where Corvettes have excelled ergonomically is in control locations. Controls should be designed and placed to require a minimum of driver scrutiny, effort and attention to operate. The first control enthusiasts

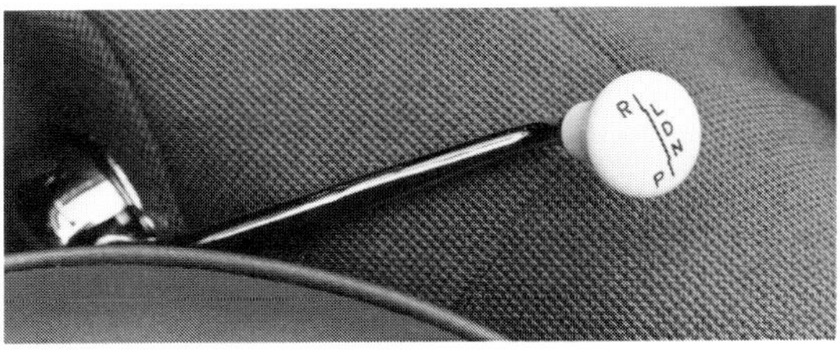

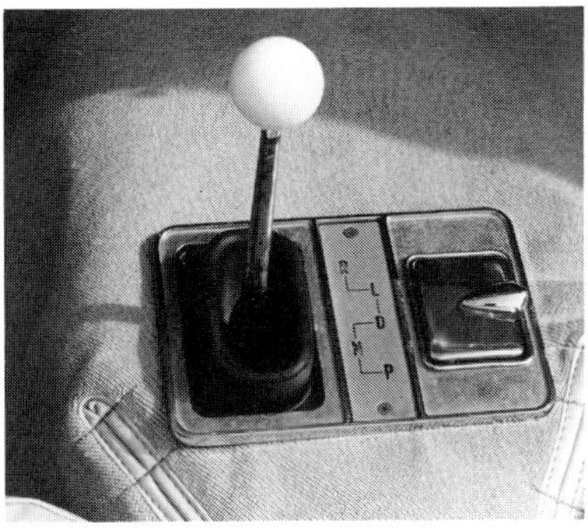

Top left: The very first Corvettes (1954 shown) had their automatic shift levers on the transmission tunnel, starting a trend for all Corvettes to follow.
Bottom left, center, and right: Whether automatic or manual (automatic 1956, four-speed 1963 and automatic 1967 shown), all Corvette shift levers have been ideally located.

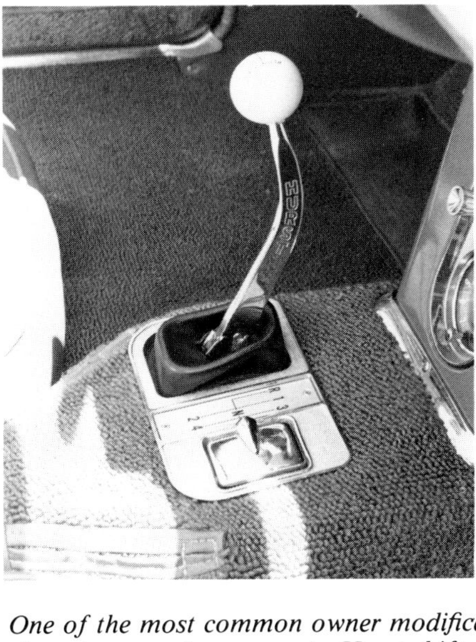

One of the most common owner modifications to 1956-62 models (1957 and 1962 shown) was a changeover to Hurst shifter linkages. No question about it, they were quicker and quieter. They're wrong for concours, but are nice non-permanent improvements for race cars or daily drivers.

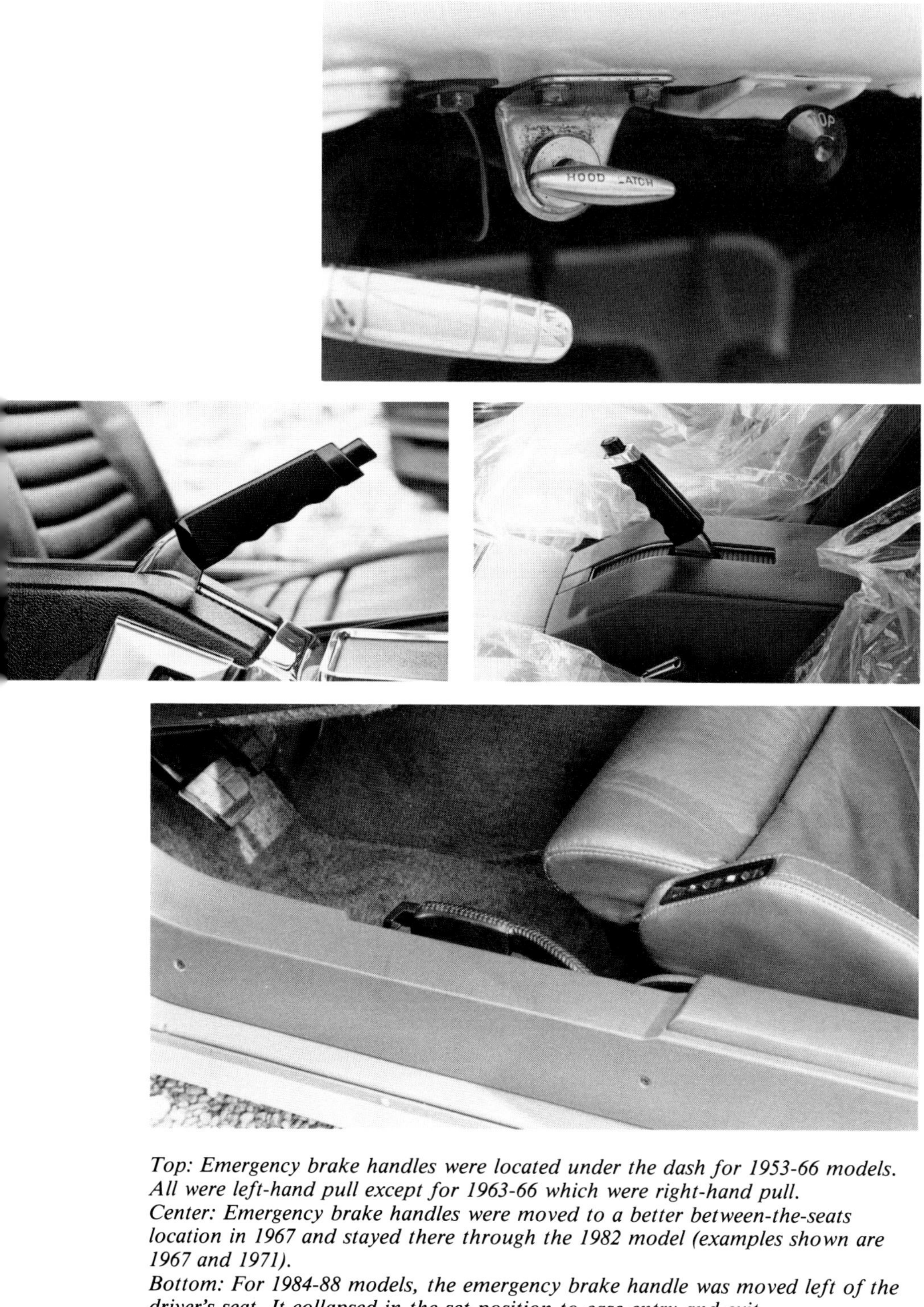

Top: Emergency brake handles were located under the dash for 1953-66 models. All were left-hand pull except for 1963-66 which were right-hand pull.
Center: Emergency brake handles were moved to a better between-the-seats location in 1967 and stayed there through the 1982 model (examples shown are 1967 and 1971).
Bottom: For 1984-88 models, the emergency brake handle was moved left of the driver's seat. It collapsed in the set position to ease entry and exit.

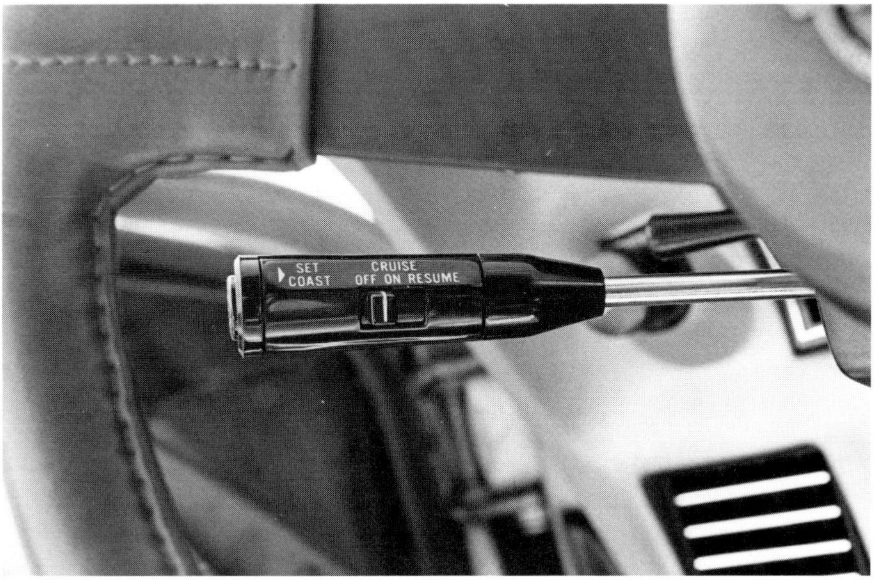

Cruise control (Chevrolet called it Speed Control) wasn't a Corvette option until 1977. Here's the control stalk for 1982 models, this one a Collector Edition.

think of is the gearshift. Automatic or manual, all Corvette shift levers from day one have been on the tunnel between the seats. There have been irritations, like no reverse lockout on four-speed shifters until 1959. Then when the reverse lockout was added, its buzzing could drive an owner nuts. But with the possible exception of the small automatic shift levers of 1953-55 models, Corvette shifters have always seemed to be in perfect locations, precisely where the driver's right hand wanted to be.

Emergency brake control for Corvettes started as a pull handle under the instrument panel in 1953 and stayed there until the 1967 model when it changed to a pull lever between the seats. It remained a pull lever to this day, except the location changed in 1984 to the left of the driver seat. The 1967-88 pull-lever style permits better leverage than the under-dash type; the location of the 1967-82 handles between the seats was the easiest to reach.

Other Corvette control locations—lights, ventilation, air conditioning, radio, turn signals, horn—look good in retrospect. There have been little modifications and corrections, but nothing to cause a buyer to select one model over another, or an owner to consider modifications.

Visibility

One last ergonomic consideration meriting mention is visibility. The 1953-62 Corvettes had wraparound windshields common to that era. The distortion in the corner curvatures could be bothersome. The 1953, 1954 and 1955 models had plastic side curtains which, in their best condition, weren't as transparent as glass. Rear vision in all the 1953-62 models was poor with the convertible tops raised, especially in the 1953-54 (and early 1955s) because the rearview mirror had no height adjustment. On the other

Left: Removable hardtops for 1956-62 models had lots of rear plexiglass for good visibility.
Right: Small window in 1956-62 convertible tops hurt visibility, but the 1953-55 models were even worse.

Top: It's one of the all-time favorite Corvettes, but the split in the "split window coupe" of 1963 blocked rear vision severely.
Bottom: The Ferrari Dino inspired the 1968-77 Corvette T-top coupe rear window treatment. Like the 1963 split window, seeing it from the outside was obviously a more important design criteria than seeing out from the inside.

hand, the removable hardtops available in 1956-1962 offered good rear visibility, except for tall drivers. Additionally, the 1953-1962 models featured a high, over-the-frame seating position that many owners really enjoy.

The wraparound windshield was gone in 1963 Corvettes, but the split rear window was in for the new coupe models. It was one of the most beautiful Corvette exteriors ever, but rear visibility obviously suffered.

The split was gone in 1964, giving an unobstructed view through the inside mirror, but the midyear coupes all had blind spots created by their thick roof pillars. The same was true of the convertible tops. The removable hardtops were better than the soft tops because their rear windows were larger, but they weren't as good as the 1956-62 hardtops (again, except for tall drivers), because they didn't have the additional side windows of the earlier models.

The 1968 model ushered in the T-top coupe (twin, removable top panels) era for Corvettes and these featured a Ferrari Dino-like rear window treatment with terrible rear vision. The 1968 design had a number of other elements—higher doors, higher center console, thick roof pillars—that created a cavelike feeling in the cockpit that bothered some owners. In 1978, the rear styling was redone to incorporate a big rear window and thinner roof pillars, reducing the closed-in feeling and improving rear vision significantly. For visibility, the convertible top and removable hardtop versions of these were very similar to the 1963-67 models described earlier.

The 1984 and newer models had high doors, high center consoles and a high entry ledge, so some of the "crawl down into it" feeling was still present, but visibility in the coupes was quite good thanks to a big rear window and relatively thin roof pillars. Vision with the convertible top up was typical of those already discussed.

By now it's clear that the ergonomic driving environment in Corvettes hasn't seen a constant improvement over the years as one might expect. In fact, it has been a series of peaks and valleys. The strengths of older Corvettes by some ergonomic measures in part explains why some owners can enjoy driving these cars daily.

2
Comfort and convenience

The previous chapter dealt with ergonomics. For automobiles, that means creating an optimum "working" environment for the driver. This chapter deals with comfort and convenience. The two topics are related, but not the same.

When modern sports cars were introduced to the United States shortly after World War II, comfort and convenience features were minimal. It was as if performance and comfort were mutually exclusive. If a car was comfortable, it couldn't be a *real* sports car.

It didn't have to be this way, but a few factors nearly dictated it. The buyer for sports cars tended to be young, so low cost was important. Low feature content meant low weight, and low weight contributed greatly to the enhanced performance of early sports cars. And young owners were simply willing to sacrifice creature comforts, if that was the price, for the enhanced acceleration, braking and cornering sports cars offered over conventional family sedans.

The Corvette arrived in this atmosphere, but General Motors couldn't bring itself to sell a bare-bones performance car. So the first Corvettes were contradictions. They weren't sports cars, but had sports car features. They weren't family sedans, but were more comfortable than traditional sports cars. When sports car enthusiasts initially rejected the Corvette, Chevrolet executives essentially said, "Well, we didn't intend it to be a sports car. It's a sporty two-seat luxury car."

Chevrolet had a nebulous vision of its first Corvette, sort of something a successful exec would toss his golf clubs into Saturday morning and cruise over to the country club. Forward-thinkers, like Corvette Chief Engineer Zora Arkus-Duntov, managed to get the Corvette on a proper performance path. Comfort and convenience were secondary for several years, then the situation reversed during the seventies when performance became socially irresponsible.

Performance came back to Corvettes with a vengeance in the 1984 model. As the design was refined in subsequent years, these Corvettes came to offer one of the best balances of ultrahigh performance and decent comfort of any of the world's premier sports cars.

Does one have to own a newer Corvette to get reasonable comfort? And does comfort automatically come with newer models? A qualified no to both questions.

By most measures, Corvettes have become more comfortable over the years. But it's more a matter of features than the intrinsic design of different

The 1953-62 Corvettes shared a straightforward chassis design that featured front coil springs, rear leaf springs and a non-independent rear suspension. Somewhat firm in stock configuration, many of these models have had even stiffer springs installed for better handling at the expense of ride quality. John Amgwert photo.

Corvette generations. Let's consider the comfort and convenience advantages and disadvantages for each of the Corvette model groups.

Ride Quality

When most people think of Corvettes as uncomfortable, they're thinking of the ride quality. There is a perception that Corvettes sacrifice ride quality for handling, and that an owner had best be prepared to suffer. It's a bad rap, a stigma that's become attached to all Corvettes thanks to the ride quality shortcomings of a minority.

The first Corvette series, the 1953-62 models, borrowed somewhat from 1949 Chevrolet passenger car chassis design, which used coil springs in front and leaf springs in the rear. These Corvettes do feel hard and choppy, but the stock springs were the best ride-handling compromise. The fact was that the suspension in these Corvettes was unsophisticated in its day, and it's downright antiquated now. It needed stiff springing to maintain respectable cornering stability and to keep from bottoming out.

If ride quality is a concern, avoid the heavy-duty springs Chevrolet made available for these models. Because the suspension didn't respond well to cornering, racers found the best solution to be minimal suspension travel, so Chevrolet obliged in 1957 and 1958 with optional front coil spring rates thirteen percent higher than stock, and rear leaf rates nine percent higher. The optional heavy-duty suspension for 1959 featured front springs eighty-three percent stiffer and rears twenty-two percent. These springs were available optionally and over-the-counter. Many have wound up on otherwise street-appearing Corvettes. They did the job on the racetrack, but they're brutal for casual driving.

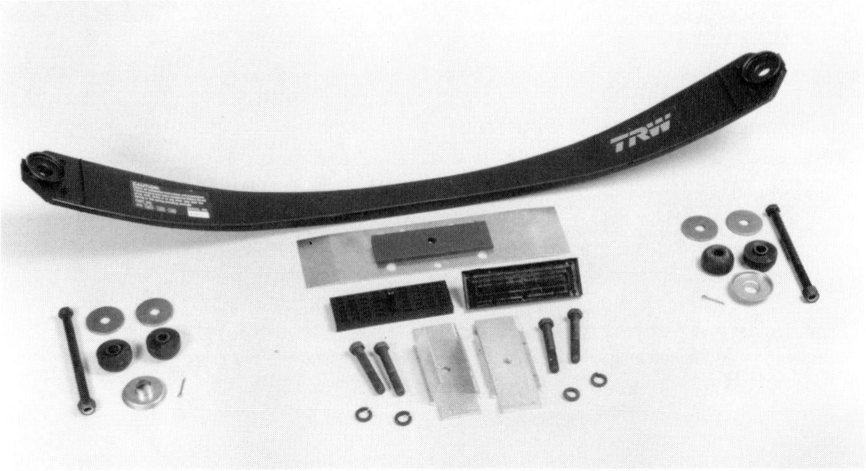

Chevrolet introduced a fiberglass monoleaf rear spring on a limited basis in 1981. It worked so well that all Corvettes since 1984 have used single fiberglass monoleafs front and rear. Shown above is a replacement fiberglass monoleaf rear spring for 1963-77 models, made by TRW and available from Eckler's Corvette Parts. It's said to improve ride quality, reduce noise and never wear out. Weight is just eight pounds. Eckler's photo.

It was a new ballgame for 1963. The Corvette chassis was redesigned to include an independent rear suspension with a single transverse leaf spring. Because of the sophistication of the new chassis, handling was substantially improved over 1962, yet the spring rates in the 1963 were lower, so ride quality was up.

Lower with the stock suspension, that is. An RPO-Z06 "special performance" option for 1963 included springs with roughly *double* the stiffness of the standard setup. Chevrolet's game plan was to offer an optimum street Corvette and an optimum racing Corvette in two separate packages.

Over the five years of Sting Ray production, incremental improvements in ride quality were made. In 1964, for instance, variable-rate springs, both front and rear, were introduced. These were softer in the short jounce mode encountered in normal driving, harder in the long jounce mode brought on by serious cornering.

The arrival of the optional 396-cubic-inch "big-block" motor in 1965 brought with it an additional 150 pounds, so Corvettes with this engine got stiffer front springs. The general feeling today is that these ride slightly harder than Corvettes with small-block engines, but some owners swear it's not so.

Magazine testers couldn't agree. Speaking of the big-block, *Road & Track* said in its August 1965 issue, "It's an interesting technical exercise, building a nice big engine like the 396 and putting it in a good chassis like the Corvette, but it honestly isn't a very satisfactory car for driving in everyday traffic. It's too much of a brute for that." Yet, other magazines, like *Car Life*, claimed handling *and* ride improved with the extra weight.

Spring rates went up in 1968 for better cornering, and wheel rim width increased (from six inches to seven) to accept wider rubber. Both caused a decrease in ride quality. Rim width increased again in 1969 (to eight inches), decreasing ride quality ever so slightly more. These Corvettes and those built through 1972 were the harshest of the 1963-82 models (excepting the Z06), but they had the highest cornering capability. As performance emphasis decreased through the seventies, each Corvette model seemed to move more toward a softer, more "boulevard-style" ride.

For the owner of a Corvette in the 1963-82 range, the base suspension will offer the most satisfactory ride quality. But what about spring replacements? General Motors, like all auto companies, wants to minimize its monumental replacements parts inventory. It attempts to do this by selecting one part that will function for more applications than was originally intended. For example, if you order a new gas lid door for a 1963-67 Corvette from Chevrolet, you'll get the 1966 style. It fits. It functions. It's just cosmetically wrong for the other four years.

To minimize spring inventory for older Corvettes, Chevrolet may make available only a heavy-duty unit because it will fit and function for both light- and heavy-duty applications. For this reason, you must determine your model's original spring rate (your Chevrolet dealer parts department can advise you) and be sure a replacement unit is in the proper range. A pleasant ride-quality surprise awaits if the car had already been changed to stiffer springing.

In 1981, Chevrolet introduced a fiberglass monoleaf rear spring, initially just for Corvettes with automatic transmissions and standard suspensions. The technology advanced, and all 1984 and newer Corvettes use a single transverse fiberglass spring, both front and rear. These springs offer substantial advantages in weight savings (eight pounds compared to forty-four), in quietness because leaves aren't rubbing against each other and, it is claimed, in ride quality.

Aftermarket auto parts companies, like TRW, offer fiberglass leaf springs for 1963-82 Corvettes in several rate grades. They are not factory original, so there is the nonstock appearance factor to contend with. They're more expensive, perhaps double their steel equivalents, but user reports are very favorable. Dick Ferrando, owner of D&A Corvette, a company in Gillespie, Illinois, which specializes in "remanufacturing" older Corvettes, uses the fiberglass spring for all available applications because his customers prefer the ride quality.

The roller coaster spring rate saga continued in 1984 when Chevrolet unveiled a new Corvette with very firm stock springs, and incredibly hard springs in the optional RPO-Z51 handling package. Corvette engineers wanted their new baby to be the best-handling production sports car in the world, measured by skid pad adhesion. They had visions of dominating Showroom Stock racing. Both goals were achieved, but the price owners had to pay in ride quality, especially with the 1984 Z51, was too high. Spring rates were adjusted downward in 1985 so that base springs were much softer, and the optional RPO-Z51 about equaled the previous year's base springs. Some rates were fine-tuned again in 1986 and 1988, but stayed in the same range as 1985. By 1987, Chevrolet made the manual transmission mandatory with the stiffer suspension to emphasize the Z51's performance nature, even with its reduced spring rates. In 1987, a new suspension option,

This is Dave McLellan, Corvette chief engineer, explaining the radically new 1984 chassis design to journalists in November 1982. Less its body, it looked like an Indy racer. In the background is Robert Stemple, then Chevrolet general manager, now president of General Motors. John Amgwert photo.

RPO-Z52, combined the softer springs with hardware previously reserved for Z51-equipped Corvettes, including faster steering.

In a comprehensive June 1987 article on the three Corvette suspensions available in 1987 models, *Car and Driver* summed it up well by saying, "Given the options available, our choice for all-around use is the Z52. It provides excellent handling and a good ride, and it works on just about any road surface. On the best roads, the base car rides a bit better than the Z52, without much loss of performance; it's the right choice for cruising Hollywood Boulevard. The Z51 shouldn't be considered for everyday street use, unless you drive only on perfect pavement."

The suspensions of the newest Corvettes can be lumped into three ride quality groups. First, and all by itself, was the 1984 with RPO-Z51. Second was the 1984 base suspension and the 1985-88 models with RPO-Z51. Third was the 1985-88 models with base suspension or RPO-Z52. For ride quality, the third group is best. The second is tolerable, but too stiff for most drivers' tastes. The first, the 1984 with RPO-Z51, is, in a word, punishing.

The next chapter describes in detail what's involved in changing to softer springing in these models. In a nutshell, the springs are inter-

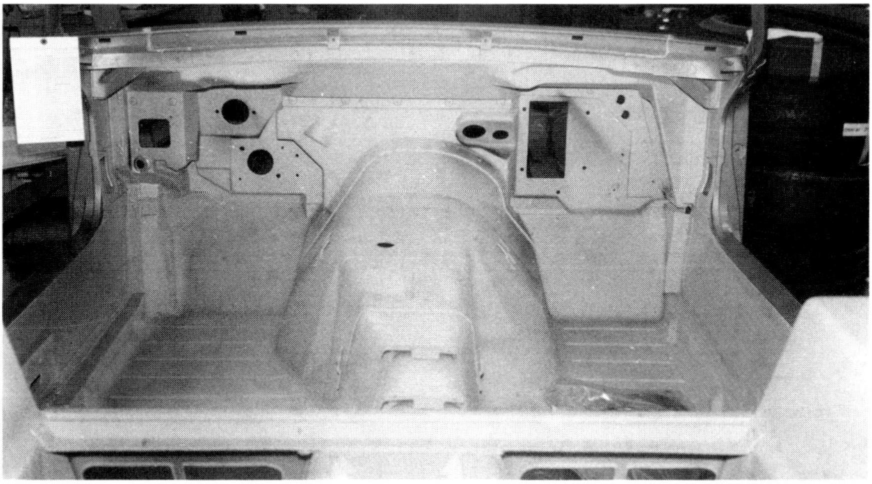

This 1984 floor pan photo shows how deeply the seating and footwell compartments were recessed to minimize overall height. Designers managed to decrease the Corvette's height without losing headroom, but the cabin is rather claustrophobic, a common Corvette complaint dating to 1968. Chevrolet photo.

---

changeable. However, to keep handling up, Corvette engineers modified the sway bars and other components in 1985-88 models to compensate for lower spring rates. Switching a 1984 to later model springs is easy and will improve the ride quality, but the handling will suffer somewhat without changing other suspension components.

Seating, headroom, cabin heat and ventilation

There's more to comfort and convenience than ride quality. The previous chapter discussed ergonomic aspects of Corvette seating, pointing out that no effective adjustments, other than fore and aft and a power base, were Corvette seat features before the 1984 model. A lack of seat adjustments certainly affects comfort as well as the ability to drive properly, so the earlier comments hold true for a comfort discussion. Additionally, there are aspects of different Corvette generations' basic design configurations which affect seating, headroom and, believe it or not, cabin heat and ventilation.

The first Corvette generation, those built between 1953-62, had a body-frame layout which put the floor pan above the frame. That was the straightforward way to do it, but it created some problems for a sleek vehicle like the Corvette. To keep the overall height low, but still maintain sufficient ground clearance under the chassis, cabin height had to be shallow. This series had its seats close to the floor to get acceptable headroom (still the least headroom of all Corvette generations, incidentally), creating a rather straight-leg driving position. The seatbacks of these Corvettes were very upright. That combination—legs straight out, back upright—is inherently uncomfortable. The seats in these models were well made and would be fairly comfortable, if they just weren't in that particular cabin environment.

Engineers rethought the Corvette for 1963 and came up with a better

In the 1953-62 period, only the plastic-side-curtain 1953-55 models had vent windows. Air conditioning was never a factory option for these Corvettes. But all did have an enormous cowl vent forward of the windshield for directing fresh air to the cabin.

Top left: Ventilation was a problem in 1963-67 coupes, so Chevrolet tried adding vents to the roof sections, including a blower motor on the driver side, but they didn't help much and were discontinued in 1966.
Top right: Last of the Corvettes to have real vent windows (actually the only ones other than 1953-55) were the 1963-67 models. Good thing, because air conditioning, though available starting in 1963, was still pretty uncommon for these Corvettes.
Bottom: Corvettes from 1968-on featured ventless side windows.

Left: Chris Condon wanted to drive his 1957 Corvette daily, so he added a Montgomery Ward air conditioner scavenged from a 1971 Ford, mounting brackets from a 1957-58 Chevy passenger car, and a clutch fan from a 1968 Chevy SS 396 for extra cooling. Except for two small holes through the firewall, no permanent body modifications were necessary.
Right: They're so well done you'd think they were factory-original, but these clever vent wings for this 1962 were hand-fashioned by an unknown previous owner.

---

idea. They formed the floor pan such that the seating areas nestled down into the frame. Like the earlier Corvettes, they routed parts of the exhaust system through, rather than under, the frame.

The 1963 Corvette accomplished what seemed impossible. Comparing convertible models, it was over two inches lower than the 1962, had a lower center of gravity, had more ground clearance, yet had over two inches more headroom in the convertible top mode. A sleeker design for the removable hardtop still yielded an inch more headroom than the earlier removable hardtops. Coupe headroom was nearly identical to the removable hardtop. Overall, the Sting Rays of 1963-67 had the most headroom of all.

Part of the headroom advantage gained with the 1963-67 models was discarded when the height of the redesigned body for 1968 dropped about two inches for appearance sake. Since the car rode on the same chassis, the decreased space came right out of the interior compartment, about an inch from headroom and another inch from the seat. This is why the seatbacks were raked an extra five degrees. The extra seatback angle was fine from a comfort standpoint, but the completely new seat's design, or lack of it, wouldn't keep occupants from submarining. So, ergonomically it was poor.

For the third major Corvette chassis configuration introduced as the 1984 model, engineers refined the clever packaging started in 1963 even further. This time they started with the floor pan under the seats as the lowest part of the vehicle, put the exhaust system through an enlarged center driveline tunnel, then literally designed the chassis to fit. This permitted another inch reduction of overall height, while still maintaining the headroom of the previous year.

It's beginning to look as if 1953-62 Corvettes win the ho-hum construction award hands down. But there is a plus side to the simplicity of their body-over-frame design. These had the coolest interiors. Why? Because much of the engine and exhaust heat was blown clear. When the engineers settled the seating compartment down into the frame, engine and exhaust heat enveloped the interior.

This unwanted cabin heat isn't a big concern in newer models with air conditioning. However, though air conditioning became optional in 1963, it didn't exceed a fifty-percent installation rate until 1971, and wasn't standard until 1980.

Fresh-air ventilation was an important design feature of 1953-62 Corvettes since air conditioning wasn't even an option. They are by far the best Corvettes in this regard. The big pop-up cowl vent in front of the windshield grabbed enormous amounts of fresh air for the cabin. The 1953-55 models even had vent windows.

The 1963-67 Corvettes had vent windows too, but they weren't enough to handle the cabin ventilation problems of these cars if they didn't have air conditioning, especially the coupes. A vent and blower were added to the driver-side roof pillars of 1964 and 1965 coupes, but the effect was negligible; they were dropped for 1966.

Vent windows were gone for good in 1968. New for 1968-75 Corvettes without air conditioning was Astro Ventilation, a pretty effective fresh-air system that pulled air into the cabin from the plenum forward of the windshield, then exhausted it through vents behind the rear window.

Chevrolet stepped up its efforts to better insulate the Corvette's cabin from unwanted heat, but the seventies also brought higher and higher engine temperatures because of fuel and emission control measures and, in 1975, those cherished catalytic converters that retained so much heat that the National Park Service warned tourists not to park over dry leaves. Be aware that any Corvette built after 1963 without air conditioning, especially the closed models, could be uncomfortably warm.

Noise

This would be a good time to talk about noise levels in Corvettes, because engineers' efforts to isolate heat from cockpits during the seventies went hand in hand with noise reduction. In some cases, one accomplished the other. Corvettes built before the seventies permitted a fair amount of drivetrain noise in the cockpit. Convertible models were, well, convertibles. Coupes were quieter; the later the better. The 1984 and newer Corvettes had, by sports car standards, very good noise isolation.

Special note must be made of the side-exhaust systems available in 1965-67 and in 1969. These had no mufflers, just chambered pipe. They're beautiful, they improve access to undercar components, they help performance, they get heat out from under the car (to where they can burn legs) for a cooler cabin, but they are loud. Some owners just can't stand them.

Legroom

Headroom has been mentioned, but legroom hasn't. This may come as a surprise, but the 1953-55 Corvettes were the worst with just forty-one inches. The 1956-62 were the best with over forty-six inches. The latter were

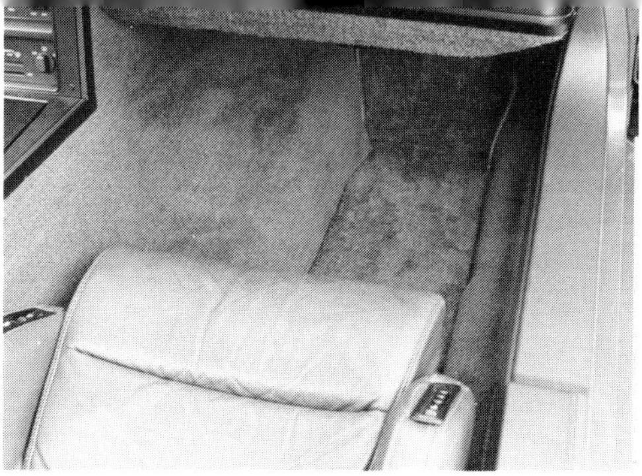

Top: Passenger legroom has generally been adequate in Corvettes, but the 1984-88 models were the exception. The huge central tunnel restricted passenger floor width severely. Depth was not a problem.
Bottom: Interior noise has been a Corvette shortcoming, particularly in pre-1984 models. Of special note were the side exhaust systems available in 1965-67 and 1969. They were loud and presented some leg-burning potential during entry and exit. They sure looked dramatic, though.

---

great cars for the tall . . . provided the soft top is folded away. Don't forget, these had the most legroom, but the *least* headroom. The legroom in all other Corvettes, 1963 to present, didn't vary a half inch.

Luggage space

Luggage space did vary. Here the coupes of 1963-67 shone with over ten cubic feet. The coupes of 1968-77 had 6.1 cubic feet, but that was increased to nearly 8.5 cubic feet by the new rear window design of the 1978-82 models. The coupes of 1984-88 dropped back to just over 6.5.

For convertibles, the 1953-62 had 4.5 (slightly deceiving due to the AMA "luggage set" measuring criteria). The 1963-67 convertibles had 8.1 cubic feet, the 1968-75 5.0, and the 1986-88 6.6. These figures were all with the tops in the up position. For 1963 and newer convertibles, the tops folded into the luggage compartment, so luggage space would be adversely affected.

Access

Speaking of luggage room, ease of access to storage and other areas varied with different Corvette models. The first Corvette series, the 1953-62

The 1953-62 Corvettes had real trunks and the spare tires hid under a flush-fitting wood cover. From 1963-on, spare tires have resided in a drop-down, rear compartment. Starting in 1984, the spares have been the undersize, temporary type. In case of a flat, how does a newer Corvette's massive tire and wheel fit into that little compartment? The compartment ratchets down into a lower, temporary position for hauling the flat off to the repair shop.

models, all had conventional, externally opening trunks. The spare tire was located in a recess of the trunk and covered with a wood panel, resulting in a flat trunk floor. Ease of trunk access is one reason some Corvette lovers won't have a Corvette newer than a 1962. The spare tire location was also preferred by many because it was easy and convenient to reach, unless the trunk was loaded with luggage or other material that had to be removed first.

The luggage compartments of Corvette coupes built from 1963 through 1982, except the 1982 Collector Edition, had no external access. They had to be reached from inside the car, usually requiring a seatback to be leaned forward. The 1982 Collector Edition, and 1984 and newer Corvettes, had external access to luggage compartments through forward-hinging rear hatch windows, and internal access behind the seats from the interior. The high lift height made them less convenient than the Corvettes with standard trunks, but better than the 1963-82 models with interior access only.

Battery access for different models has varied as well. The 1953-62 Corvettes had a conventional and accessible underhood location. The 1963-67 models used an underhood location also, but the batteries of some

Conventional trunks of 1953-62 Corvettes offer easy and handy access to luggage and whatever, unless the spare tire must be reached with the trunk fully loaded. Access to later Corvette models (1967 convertible and 1985 coupe shown) is less convenient.

---

were tough to remove. With non-air-conditioned small-blocks, the radiator expansion tank had to be loosened and moved out of the way. With the 396-cubic-inch engine of 1965 and all air-conditioned midyears, the battery came out through a removable wheel house panel after the left front wheel was removed. The non-air 1966-67 big-blocks were the least trouble since hoses just had to be shoved out of the way.

Top: Batteries of 1953-62 Corvettes were under the hood just forward of the firewall on the passenger side.
Center: Batteries of 1963-67 Corvettes with small-block motors and no air conditioning were in the same approximate location as the earlier years, but removal was more difficult because of heater hoses and the radiator expansion tank.
Bottom: Air conditioned 1963-67 models and the 1965 with 396 engine had driver-side battery locations. The batteries were removed through a panel in the wheelhouse.

Access to almost everything in the engine compartment of 1984-88 models was excellent, thanks to the forward-hinging clamshell hood. Battery removal, however, required moving the optional cruise control unit out of the way on 1985-88 Corvettes.

Left: The wraparound windshield of 1953-62 Corvettes (and the steering wheel of 1953-55s) created an entry obstacle in an otherwise easy configuration to get in and out of.
Right: A handy feature 1953-62 owners like is the vertical glovebox between the seats.

Top: Corvette radios started in the center dash area, then moved to the center console in 1958 where they've been ever since. Radio choices were quite restricted until 1977 (1982 style shown at right) when the console was made larger. Bottom: If your stock Corvette radio isn't up to snuff, take heart. High quality replacements that fit perfectly into existing radio spaces are available. This one fits 1968-76 Corvettes, is made by Pioneer, and sold by Mid America Corvette Supplies. Mid America photo.

Corvette engineers were in a no-compromise mood with the 1968 model, so they moved the battery to an unhandy location behind the seats for better weight distribution. Also, moving the battery helped alleviate some of the growing congestion in Corvette engine bays. Access to the battery was clean, but jump starting it meant laying cables over interior components, and removal was awkward. All but the 1968 and early 1969 batteries were side-post style, so the conventional top-post battery strap won't work. It's not a bad idea to lay some kind of strap under the battery before replacing it to make the job easier the next time around. The battery remained in this behind-seat location through 1982.

Batteries moved back to the engine bay in 1984. It has to be tilted around a body support brace for removal in 1984s; the cruise control unit has to be moved in later models. Generally speaking, access to everything in the busy engine bays of 1984 and newer model Corvettes is greatly enhanced by the huge, forward-tilting, clam-shell hood.

Radios

One last convenience item worth a mention is radio selection. For the 1953-67 models, radios were take-it-or-leave-it, except for 1953-55 when a radio was a required option. There were no choices of radio types. In the 1953 through early 1963 models, the radio was a signal-seeking, pushbutton AM style, a premium model for its day. In mid-1963, a pushbutton AM-FM (not signal-seeking) was offered. These weren't stereos; they came in 1968 when customers had the choice of AM-FM mono or AM-FM stereo. This continued through 1976.

Radio selection was limited in Corvettes built through 1976 because the Corvette's limited console space availability required a shallow, unique radio. In 1977, the console was redesigned with more depth to permit a wider selection of standard Delco radios. In 1977, an AM-FM with stereo tape could be ordered along with the AM-FM mono or AM-FM stereo. In 1978, yet another possibility was added, an AM-FM stereo with CB (citizens' band). The 1977 and 1978 tape units were 8-track. In 1979-82, either 8-track or cassette units were available.

The 8-tracks were gone in 1984, but an ultrapremium Delco-Bose AM-FM stereo with cassette hit the option sheets at $895, undeniably the finest factory sound system available up to that time. A non-Bose Delco with AM-FM and cassette was also available, as was a CB unit. All were electronically tuned.

The CB continued as a radio option in 1985, but was dropped for 1986 as sales for 1985 were only sixteen units.

Radio quality is one Corvette convenience feature that has improved over the years without ever regressing. The sound systems in new Corvettes are a world apart from those used in earlier Corvettes. Anyone can drop a high-quality radio in any model automobile, but aftermarket manufacturers have recognized that owners of early Corvettes interested in upgrading radios would want to do so without any damaging modifications. As a result, upgraded radios are available for earlier Corvettes which are perfect fits and, in some cases, perfect visual matches. And the selection is improving.

So it seems the perception of Corvettes improving steadily in comfort and convenience is only partly true. Some individual comfort and convenience features have gotten better, but others have bounced up and down over the Corvette's life span. No one Corvette model or series can claim the most-comfortable title, but with a little owner help, many can be very pleasant long-distance runners.

3
Underpinings

A Corvette's underpinings are its chassis and, more specifically, its suspension. One goal here is to explain the positives and negatives of different Corvette models with regard to different handling characteristics. Another is to review some things drivers can do, with and without hardware modifications, to compensate for a Corvette's shortcomings and to make the most of a Corvette's strengths.

What goes on under a Corvette is a mystery to most drivers and usually not much fun to read and learn about either. I thought the best approach for presenting this subject would be to include comments and advice dialog from someone with impeccable Corvette suspension and driving credentials. But who?

Dick Guldstrand. He is among the most celebrated of all Corvette race drivers, yet his reputation as a Corvette suspension wizard is even more recognized.

Dick Guldstrand built his own car two years before he was old enough to drive it legally. He raced professionally on dirt tracks for fun while he was in high school, and for extra spending money while he was earning his engineering degree from the University of California at Los Angeles.

Corvette racing for Guldstrand started as a spare-time hobby in 1959. The car was his personal 1957 Corvette. By 1961, he was racing practically every weekend. Driving new Sting Rays, Dick Guldstrand won Pacific Coast Sports Car Club of America (SCCA) championships in 1963, 1964 and 1965. He was voted the California Sports Car Club Driver of the Year in 1964. He drove the awesome Corvette L88 to the GT class victory at Daytona in 1966. He drove one of the legendary Grand Sport Corvettes for Roger Penske at Sebring in 1966. And on and on.

Dick Guldstrand phased out of active racing competition in the seventies, but phased into the business of providing suspension services and parts, mostly for Corvettes, through his company, Guldstrand Engineering. The transition was natural and easy because every car Guldstrand raced had a Guldstrand-prepared suspension.

Guldstrand was fortunate in where and when he grew up. His home was, and still is, Culver City, California, right in the area where hot-rodding was born. As a kid, he knew and worked for, or around, the men who started America's performance aftermarket business, men like Vic Edelbrock, Ed Iskenderian, Stewart Hilborn, Lou Senter, Dean Moon and Ted Halibrand.

I asked Dick when he first became involved with Corvettes, and he said he fell in love with the Corvette the first time he saw a 1953 model. But because of school, marriage and his job, he couldn't get serious about racing until 1959.

When Corvette greats like Dick Guldstrand raced their early Corvettes, rules permitted wheels a half-inch wider than stock. A wheel and tire change is still a popular non-permanent modification to alter handling and appearance for many models. There's certainly no shortage of aftermarket wheels and tires available.

"I raced my '57 Corvette in 1959 and 1960 on my own," Guldstrand remembered. "Then, in 1961, I picked up a sponsor, Baher Chevrolet in Manhattan Beach. So I sort of phased out of my engineering job and into racing because I was doing quite well. I was really excited with how sports car racing was coming up. In fact, in 1962 Baher Chevrolet hired me as their performance manager, in addition to sponsoring my car."

While we were on the subject of his 1957 Corvette, I wanted Dick's opinion of its suspension's strong and weak points.

"Well, it was basically a '49 Chevy. Looking back, it was state of the art . . . then. But I cringe when I think that we raced them. It was grim. It would flop and wallow. It wouldn't go straight. I had to make special front springs for it. The geometry was so bad—not that I was complaining about it; I didn't know any better—that we had to make a go-kart of it. We couldn't let it move a lot because of the gyrations it would go through."

Were these inherent characteristics of the car?

"Sure," Guldstrand maintained. "In the front, you couldn't get any caster gain or camber gain out of the way it was designed. And you couldn't change the physical location of the parts. So what did you do? You made a go-kart out of it and slid it around corners."

Viewed from the side, caster is the angle created by a line between the upper and lower steering pivots and a vertical line. Increasing caster tends to make steering heavier at low speed, but more stable at high speed. Viewed from the front, camber is the angle a tire makes with an imaginary vertical line. Positive camber tilts the top of the wheel outward; negative tilts it in-

Here's Dick Guldstrand sliding his 1957 Corvette around corners at Santa Barbara and Riverside raceways. According to Guldstrand, the suspensions of the early Corvettes went through such violent geometry changes during cornering that very stiff spring rates had to be used to minimize suspension movement. "We had to make a go-kart out of it," said Guldstrand. Once the suspension travel was limited, however, the right driving technique made these Corvettes unbeatable in their day. Dave Friedman photos.

---

ward. Dick's answer implied that the Corvette's suspension design didn't permit the caster and camber to be controlled through the range of chassis movement encountered in racing. I wanted to make sure I understood his solution was to stiffen the chassis.

Guldstrand confirmed, "Oh, did we ever. Two sway bars and railroad-car coil springs up front . . . tied down the back. We stiffened them so much they were just go-karts; that's not a very high-tech description, but that's what they were."

But did it work?

Again, Guldstrand confirmed, "You bet. In fact, once you stopped most of the suspension travel, the car was incredibly predictable. If you look at old photos of those cars racing, everybody is sideways in the corners. With my experience in midgets on dirt tracks, this was very natural for me. Some of the less experienced drivers tried to *drive* through the corners. You couldn't do that. You had to slide through the corners. I *threw* it into corners and got crossways, just put on a hell of a show. But it was done for a reason. We couldn't tolerate the dramatic geometry changes that came along with a lot of ride travel. Some of the more sophisticated cars like the Maserati, Porsche and others could tolerate some ride travel, but we couldn't with the Corvette."

We talked about the "heavy-duty" suspension parts Chevrolet offered as options and across-the-counter, Dick mentioning that what Chevrolet offered was a direct result of what racers learned on the track.

Asked if he would recommend suspension modifications to an owner of a 1962 or older solid-axle Corvette, who intended to use the car only occasionally and gently, Guldstrand replied, "No, I wouldn't touch it. Leave it just the way Chevy built it."

What about the owner who wants to do some spirited street driving or weekend slaloming?

"A wider wheel helps," Guldstrand continued. "And of course, a good performance tire is a must. But you've got to come up in spring rates. That's the one thing that'll help that car more than anything. I'd recommend an increase of at least twenty or thirty percent in rate over stock springs. I went even more; I went fifty percent or more.

"Come up the same percentage front and rear. What you've got to control is the roll. If you've ever seen one of these go around a corner at speed in stock configuration, it looks like it's rolling over on the door handle. I mean, it's six or seven degrees of roll. It's bloody awesome. Of course, when you do that, the tire rolls under . . . it's almost undriveable.

"So you've got to control the roll with higher spring rates. Ideally, when you change springs, you should also get the car down to lower the center of gravity.

"Now, in order to control all this, you'll obviously need a better shock absorber and more precise bushings. Plus, and this is very important, anytime you're going to do a car, the basics have to be spot-on. The kingpins and bushings, the tie-rod ends, the steering bearing . . . all of these have to be perfect. Remember, we were working with fairly new cars then. I guarantee you that by now much of this is just about worn out. These cars were so durable and wore out so slowly that owners just don't realize it. The problem surfaces when you start to demand more of the vehicle. The basics have to be right on before you start changing anything. That's true for any Corvette."

This all presents a dilemma for the owner who wants more out of his car but is reluctant to change it. Would an upgrade to the factory heavy-duty springs make sense?

Guldstrand felt it would. In his words, "Sure, that would be ideal if the parts could be rounded up. The one thing an owner has to decide to begin with is what the ultimate goal of the vehicle is. You don't want to be changing horses in the middle of the stream. If you want an original car, don't mess with it. Get it right back to original, right back to the right part numbers, put it together and just drive it as it is. They're so valuable today, it almost makes me cringe to recommend changing it.

"But if an owner wants to slalom or go vintage racing, he shouldn't go out and embarrass himself with an absolutely stock Corvette. The heavy springs, shocks and other things we mentioned are a must to go fast."

Dick deemed wheels and tires very important. "When I was racing these cars, the rules said you could use a rim that was a half-inch wider than stock, so everyone did, and that's still good advice," Guldstrand recalled. "That allowed us to run a little wider tire. By the way, only treaded tires should be used. I'd never recommend running a slick on that car. The lateral force you can generate with a slick is absolutely detrimental to the overall performance of the vehicle. This was one of the things that hurt the competitiveness of the solid-axle Corvettes. When these really good slicks started showing up in the early sixties, the early Corvettes couldn't handle them; there was way too much lateral load. The car did terrible things. Go with a good, standard-treaded, high-performance tire. You can use either a radial or bias-belted tire. You can have some fun with it.

"But you have to decide in the beginning where you're going to go with it. I would stay with factory parts if originality is important. What we're finding with our own customers is that often when they find a Corvette that's so cosmetically poor that it really can't be restored, they put a roll hoop in it, gut the interior and go racing . . . have a good time. But keep it in the vintage of racing that it was intended for. To try to make it more of a racer than it was in its era doesn't make sense.

"On the other hand, if you want to restore that car, don't go racing with it. Leave it stock. You'll have a lot more fun driving it on the street. It'll be much more comfortable."

How big a factor was comfort? Could you drive one of these Corvettes on the street after it was set up for racing? Guldstrand said he had no choice in the first couple of years, that he had to drive his car to the races, but that it was awful on the street. Since this was both race and street car, I was curious about the changes made and reliability.

Guldstrand remembered, "We made a wraparound seat, kind of like those used in old sprint cars. You could gut them a little bit. You could take the carpets out. You couldn't take the door panels out or change the dash and all that until much later. You definitely had to have a roll hoop in it. You could take the windshield off and replace it with a little plexiglass screen. You had to get the metallic brakes, and we did a little extra brake cooling by drilling the backing plates and ducting some air to the drums.

"Those cars were very reliable. The only trouble we had initially was breaking some axles. That was really dumb. There was a little ring that held the bearing on. If that little ring pulled out, which it did, then the whole axle would come out. So we had to tack-weld the ring in place. But the Positraction rear end, the mechanics of the whole chassis, were excellent. There was nothing to fear in the design of the components."

The discussion shifted to the 1963 Sting Ray. Dick, unlike nearly all the other California Corvette racers, selected a convertible instead of a coupe for racing, because he felt the open car's lower center of gravity would more than offset the coupe's aerodynamic advantages.

Asked to contrast the 1963 suspension with that of earlier Corvettes, Guldstrand said, "Well, the single biggest change they made was the independent rear suspension. That was a real breakthrough, because it took all the weight out of the unsprung element in the rear. That's what made the earlier Corvettes hop and dance so badly.

"The design of the '63 rear suspension was really beautiful. However,

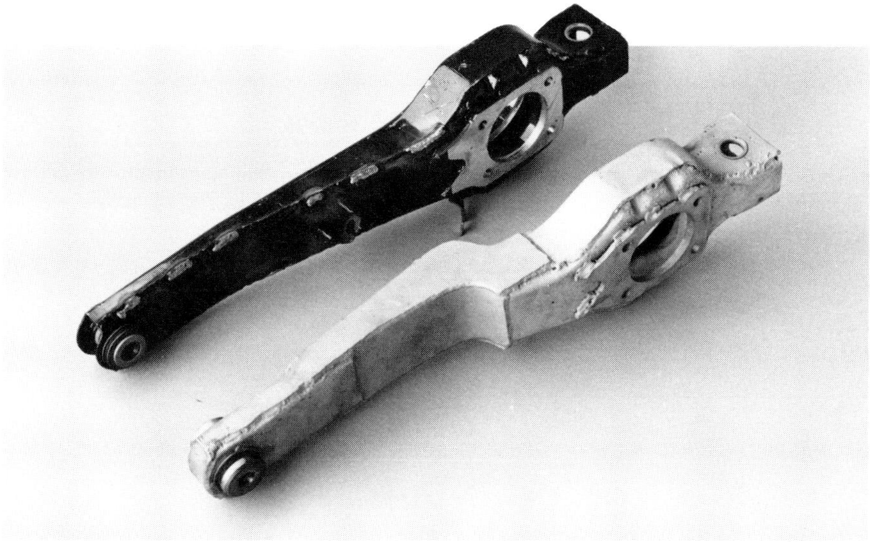

The 1963-82 Corvette independent rear suspension employs a trailing arm design. Guldstrand Engineering sells the modified trailing arms shown. They increase clearance between the arm and tire sidewall. They're stronger than stock and require less wheel offset, thereby lessening wheel bearing loads. One style uses stock rubber bushings; a racing version uses metal spherical joints. Guldstrand Engineering photo.

---

it did have a great deal of roll understeer. The designers had the option to make it either roll understeer or roll oversteer."

I asked Dick to explain his last statement.

"It's a trailing arm design, and with a trailing arm you can't completely bump steer it. You can't take all the toe steer out of the rear end because of the angularity and the arc prescribed by a trailing arm. There's no way out of it. You're stuck. So given the choice, it was designed with a lot of roll understeer because there's a terror in Detroit about selling an oversteering vehicle."

Looked at from above, toe is the angle between the tires. Toe-in, for example, means the distance between the fronts of the tires is closer than the distance between the rears. Understeer is loss of adhesion in front. When Dick spoke of Detroit's terror, he meant the practice of all manufacturers designing understeer into automobiles sold to the public. This is because the vast majority of drivers will react to a loss of adhesion, front or rear, by letting off the accelerator and getting on the brake. That shifts weight forward, which tends to stabilize an understeering vehicle by restoring steering control. The same driver reaction to an oversteering vehicle could easily break the rear loose and start a spin.

Racers, however, don't like understeer, or "pushing," as they call it. They'd rather be losing adhesion at the rear, because steering is maintained. So long as they've got steering, they can "drift" through turns with the rear in a controlled state of sliding, and they can bring the rear back into full control by shifting weight back to the rear (assuming rear-wheel drive) by easing the throttle down.

While most other Corvette race drivers favored the coupe version in 1963 because of its superior aerodynamics, Dick Guldstrand chose convertibles for their lower center of gravity. Here he is at Riverside in 1963 racing against Bob Bondurant in the Cobra, and again at Riverside in 1964. The new suspension of the 1963 Corvette made it much quicker, but drivers familiar with the old solid-axle models had to change their technique. Dave Friedman photos.

Guldstrand agreed, and continued. "Well, I had to balance it. Right away, we had to lower the car and raise the front and rear spring rates forty percent. I got it controlled so that the change in geometry in the rear didn't create all this roll understeer. That's what I had to get away from."

Dick went on to say that some of what he needed was available directly from Chevrolet; other things he had to find or fabricate. He indicated great pleasure with the fuel-injected 327-cubic-inch motor, from which he was able very quickly to get 400 horsepower. He said the clutch was good, the transmission great. But what about steering?

"I had to figure out how to bump steer it," Guldstrand emphasized. "I could get the roll steer out of the front. That's something I don't think anyone else was even looking at in those days. I started by bending the steering arms up so that I got almost three quarters of an inch of roll steer out of it. I got it to where it would bump steer almost to within twenty-five or thirty thousandths. I did it with two dial indicators on a plate I stuck on the spindle . . . moved it up and down, kept measuring and fiddling until I got it. Pretty crude, but did it handle!"

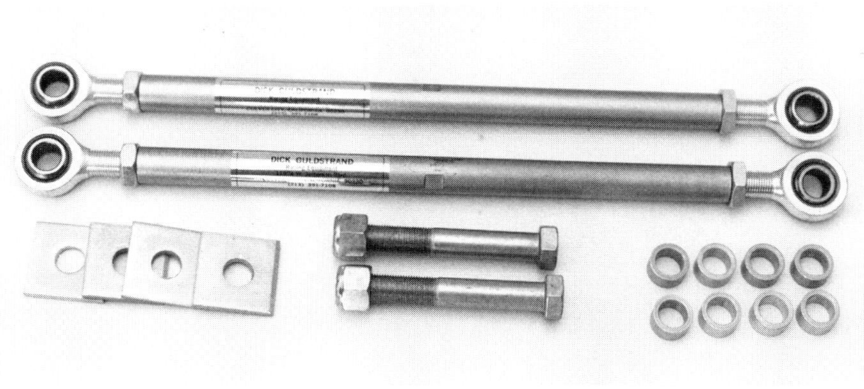

Here's Guldstrand Engineering's camber rod kit for 1963-82 Corvettes. This eliminates the lower rear strut rod rubber bushings and the rear eccentric camber adjusters. Guldstrand says it provides simple and accurate camber adjustment that stays put. Adjustment is provided by left- and right-hand threaded heim joints. Guldstrand Engineering photo.

Roll steer is toe change caused by body roll. Bump steer is toe change cause by vertical suspension travel. Racers want both minimized.

Earlier in the conversation, Dick mentioned that he'd lightened his 1963 Corvette racer. When asked how, he said the headlight motors were removed, the fiberglass bonding strips were chipped out and a lot of the structure was removed. I assumed that included the frame itself.

"No, no," Guldstrand corrected me, then continued. "We strengthened the frame because it's *low* weight. We added material there. Everybody else was worried about getting the weight out of the frame. The balance of a vehicle front to rear is incredibly important, especially with a Corvette. You wanted to get forty-eight to fifty-two, front to rear, or whatever weight you could get to the rear. So you've got to be very careful not to lighten the wrong parts of the car."

What was the 1963 Corvette like to race with compared to the 1957?

According to Guldstrand, "Well, with the 1963 you were going a hell of a lot faster. The car was working a lot better. The tires were better. The chassis was better. We'd learned a lot about the motors and were getting a lot more horsepower out of them. So now it got down to driving finesse, lines through the corners . . . your technique. You didn't just throw it into the corners, dirt-tracking it all over the place. You had to worry about precise lines and acceleration speeds.

"With the old car, when you got sideways you couldn't get back on the power. I actually got the rear too firm at first and had to soften it a bit to get the rear to hook back up again. It was power-oversteering too dramatically. I had to lighten up the rear springs a little, until I got a spring rate that really hooked up well. I built a lot of those early Corvettes for other people . . . a lot of them."

Despite the perception that the Cobra was unbeatable, Guldstrand won the SCCA regional championship in 1963, 1964 and 1965 at the expense of the Cobras.

Guldstrand explained, "See, if anybody made a mistake, I was right there. I couldn't beat the Cobras. But I would always finish second or third, or first. My finishing record was ninety-nine percent. All they had to do was bobble, and I'd beat 'em. I won my first two regional championships because no two Cobras ever finished, and I finished them all. So I won the championship right there in Shelby's backyard."

I'd asked earlier what Dick would recommend in the way of suspension modifications for the owner of a 1962 or older Corvette who wanted to do some spirited street driving. The question was repeated for the owner of a 1963-67 model.

Guldstrand replied that he has 60-series tires on his own '67, a mildly modified street car. He said the profile of today's 50- and 60-series tires permits a lot more rubber on the road without body modifications, but he did caution that the emergency brake bracket does have to be moved from the side of the trailing arm to the top. Otherwise, if anything went wrong and the tire's sidewall moved, it would go right into the bracket and puncture. Guldstrand's personal '67 also has polyurethane bushings, a different sway bar, and Monroe Formula GP shocks. Guldstrand is biased toward Monroe understandably, because his engineering company helped develop the Formula GP shock line. Here's what he said.

"My company has an engineering consulting contract with Monroe to develop Formula GP shocks for most cars. The Formula GP is a two-tube, gas-filled shock. It's really exciting to work with Monroe, an American shock company, on state-of-the-art projects. The Formula GP shocks are fantastic, and the Corvette applications are among the best. Other than tires, changing to that shock is the single best thing you can do to improve ride and handling. We've put these on customers' cars, and they swear we've reworked the whole suspension."

The conversation turned to power steering versus manual, and Dick said he has power on his '67 and much prefers it. Asked if he'd recommend owners with manual steering change to power, he said as far as he was concerned, yes.

"I'd definitely change. If you stay with manual for street use, you have to make sure the steering arms are set on the slow-ratio setting. We always used the fast ratio for racing, but it's too hard to steer at slower speeds with manual steering. You'd never get it parked."

Dick also mentioned that his own '67 has had the steering box blueprinted and, though he's using standard spring rates, he put in a fiberglass rear leaf spring because his wife drives the car a lot, and he believes the plastic spring gives a smoother ride.

He wanted to emphasize one other point. "For most people, I would highly recommend that they do nothing to their Corvette that isn't reversible. In my shop, we do almost no cutting or welding to customers' cars. Our modifications are bolt-on.

"Some guys weld the cross-member in the back to hold the rear end in. Don't do that for a street car. For a race car, fine. It's the only way to go for racing. I took all the rubber out of my race car. It looked like rubber, but it was black phenolic. It was extremely stable. In fact, the '68 L88 Corvettes we did for James Garner were the fastest GT cars they ever had at Daytona. They'd do almost 200 miles per hour."

I commented that the suspension introduced in 1963 stayed much the

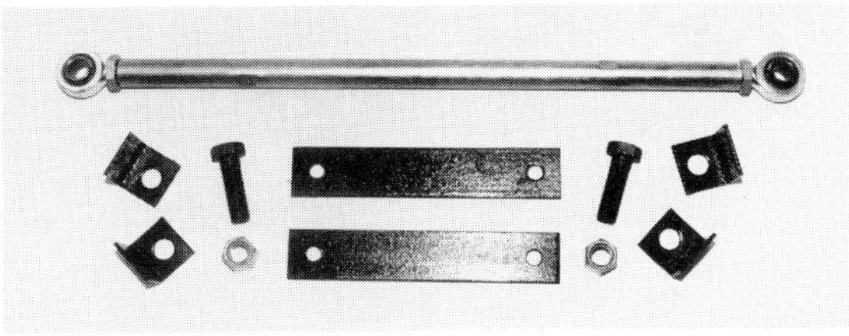

Left: Guldstrand Engineering offers front coil springs for 1963-82 Corvettes in stock, slalom or race spring rates. Guldstrand Engineering photo.
Top right: These control arm bushings for 1963-82 models are available from Guldstrand Engineering in metal (with grease fittings) for serious racing, and polyurethane for serious street or slalom use. Guldstrand Engineering photo.
Bottom right: Guldstrand Engineering's spreader bar kit reinforces the 1963-82 chassis by connecting the front suspension towers with an adjustable heim-jointed bar to eliminate chassis flex and unwanted camber changes. Guldstrand Engineering photo.

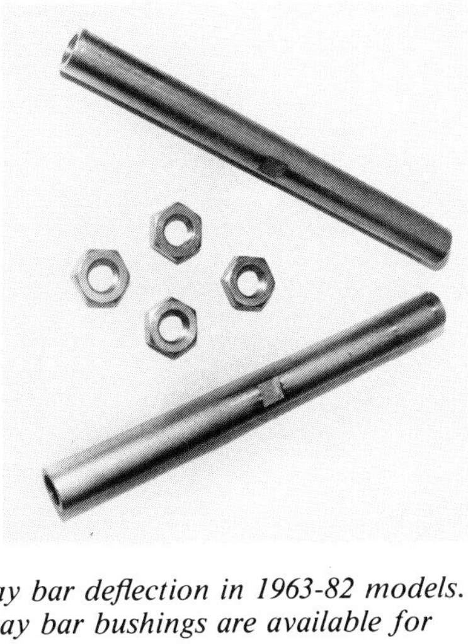

Left: Aluminum sway bar blocks eliminate sway bar deflection in 1963-82 models. They're for serious race duty. Polyurethane sway bar bushings are available for all Corvettes built after 1963 and are for use in all but serious race duty. Guldstrand Engineering photo.
Right: These Guldstrand heavy-duty steering rods replace the stock stamped rods of 1963-82 Corvettes and use existing tie rod ends. They eliminate flex caused by heavy loads generated by high performance tires, and provide easy toe adjustment. Guldstrand Engineering photo.

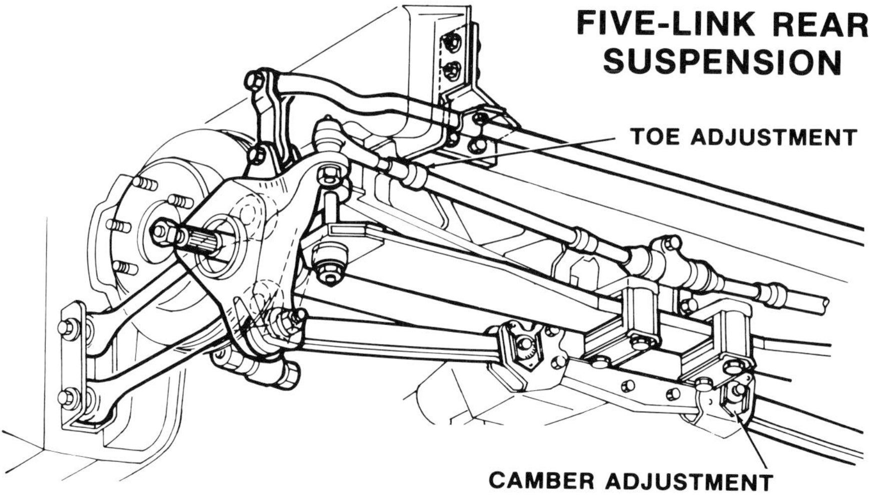

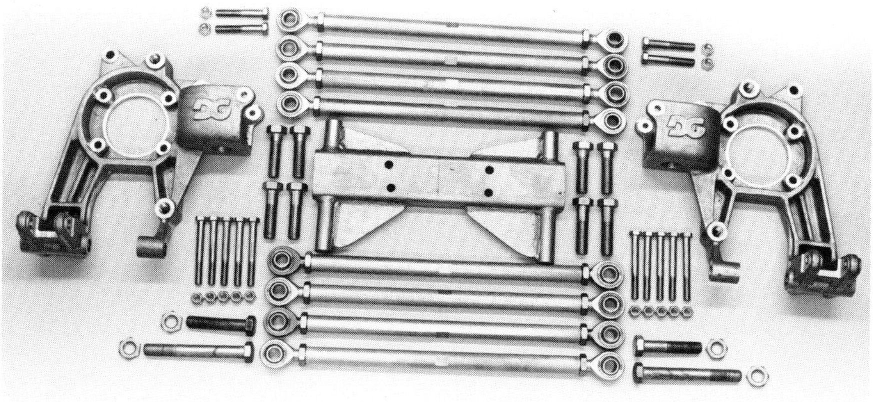

The top illustration is the five-link rear suspension Chevrolet introduced in the 1984 Corvette. The photo is of Guldstrand's five-link suspension for 1963-82 models. This is a serious, relatively expensive chassis modification requiring welding, but one that Dick Guldstrand developed and raced with successfully for years. Chevrolet illustration. Guldstrand Engineering photo.

---

same through 1982, and Guldstrand agreed, putting it this way:

"Almost identical. The wheel widths increased from six inches to seven in 1968, then to eight inches in 1969. There were some little geometry changes here and there such as bigger front spindles, bigger bearings in the front, better steering arms. They improved the rear wheel bearings. The half-shafts got bigger. The Positraction got more plates in it. The big blocks were available from 1965 through 1974, and so there were heavier springs to handle the weight. The small-block displacement increased in 1969 from 327 cubic inches to 350. Things generally improved until 1973 or 1974, then started going downhill.

"Those years in the seventies were devastating because of the gas crunch. It was terrible. By then I'd been through all these wonderful racing activities—driving the L88 and Grand Sport for Penske, driving for Honda in Japan and Europe, working with Jim Garner, doing the GTP Group Seven program, the CanAm, the TransAm, the NASCAR circuit—then all of a sudden, I thought, 'Is it over?'

"I think the one thing that Detroit forgot for a while (and it almost did them in) is the ongoing love affair between man and his machine. If he could get it in the house, he'd take it to bed. The minute you forget that and start sticking people in little bread boxes, you're not going to sell. The Japanese figured that out in a hurry. All of a sudden, you've got 240Zs and RX-7s and on and on. They're exciting cars. Meanwhile, the Corvette's going down the tube. A few stupid things like gas prices and the DOT don't change anything. It took fifty years to develop the love affair. We're not going to lose it over one gas crunch.

"We almost lost the whole market. I couldn't believe it. I could *not* believe it!"

The name of Dave McLellan, Zora Arkus-Duntov's successor as Corvette chief engineer, came up at the beginning of a discussion about the development of the 1984 Corvette.

"Can you imagine the integrity, strength and energy McLellan had to have?" Guldstrand said with obvious admiration. "You've got to remember that the program started in 1978. Duntov brought Dave out to Culver City to meet me and to be sure the enthusiasts' feelings were part of any new Corvette. Dave and I sat for hours and talked about what a Corvette should be. Mid engine? Aluminum? What is a Corvette? Boy, that was hard to define. But I felt very strongly that it should be a front-engine, state-of-the-art, exciting, macho, beautiful American plastic pachyderm. That's what would sell. All these years, all these guys have dreamed about, *dreamed* about owning a Corvette. Don't change it. Don't screw it up. Don't make it a funny little four-cylinder, mid-engine thing."

It sounded like Guldstrand had contributed to the chassis design of the 1984 Corvette, but when asked, he replied, "Not really. We were involved in some of the tire testing. See, everyone was of the opinion that the car was going to weigh around 2,700-2,800 pounds. So the brake size, the springs, everything was perfect. Everybody was happy. But the weight crept up, like 500 pounds, and then there were problems. Spring rates were run up to control it. A little-known engineer with tremendous talent, Fred Schaafsma, was in charge of that program."

Fred Schaafsma was little known at the time, but he is, at this writing, little known no longer. He's chief engineer for all Chevrolet products. But back to the 1984 Corvette. Was Guldstrand happy with the car's layout with regard to the suspension?

"Oh, yes. No complaints at all. I do think they have to control the bump steer better. It's so close to a race car. Can we have the best of both worlds? Makes me nervous. For the novice, the car still has to be very predictable and very stable.

"This car is very close to a race car. It is fast. The first time I drove the 1984 Corvette was for testing at the Willow Springs track. I was ten seconds a lap faster than my Trans-Am record in 1967! That's with the stock engine, air conditioning and stereo on, and no roll cage. That's awesome."

Talk turned to the RPO-Z51 suspension and Dick called it "terrible." But he went on to explain.

The majority of Guldstrand products have been developed for 1963-82 Corvettes, but Guldstrand has also tweaked what he admits is an already very good 1984-88 chassis. His Grand Sport 80 package for newer Corvettes features big-ticket items like blueprinted Traco Engineering motors, and details like polyurethane control arm bushings complete with grease fittings.

"It was beautifully contrived because it was a race car. But they didn't tell the dealers that. The dealers didn't understand that this was a real beat-the-Porsche 944, beat-the-Lamborghini-Countach race option. That's why the Z51 was done. Chevrolet knew it was going to take on the world with that car, with showroom stock racing and everything else going on.

"Remember, the Corvette had been in the doldrums. It didn't have the horsepower. It didn't handle. This new one had better do something right. And it did. That 1984 Z51 Corvette would outhandle anything . . . *anything*. It might rattle your cage a little bit, but it would handle!"

Dick then mentioned that his shop in California is changing the Z51 suspensions of 1984 Corvettes for a lot of customers. I asked for some details.

"We put in the base 1986-87 suspension," Guldstrand said. "Steering rack, springs, sway bars . . . the whole shot. We have to make some changes, but it goes in. It's about a $1,600 package, and it's absolutely worth it if you want to keep your wife, your kidneys, or whatever."

Is this changeover something a guy can do himself? Guldstrand said yes, provided the sway bars weren't changed.

"If you change the sway bars on the '84, you have to change the whole radiator shroud, the angle of the radiator, the way the fan goes on . . . everything has to be changed. You have to pull the whole thing out and put the later one in. But you can just change the springs, and they're direct replacements."

The reason Dick pointed out the sway bar issue is that when Chevrolet softened the spring rates on the Corvette after 1984, engineers maintained most of the handling by increasing the size of the sway bars. By switching a 1984 Z51's springs to those of a base suspension 1986-87, the ride quality will equal the new model, but the handling won't because the 1984 had smaller sway bars.

"That's exactly right," Dick went on. "And there's a little more to it. The newer suspension gives you more antisquat and other things so it doesn't nosedive with the softer springs. Subtle things like that. The 1986-87 suspension is really well done. I'm so impressed with it."

It should be pointed out that Guldstrand's comments were made in 1987, before release of the 1988 model. The 1988 Corvette had further suspension refinements, including slightly stiffer front spring rates, a choice of two new brake systems, and optional seventeen-inch wheels.

Back to our conversation with Dick. The RPO-Z51 came with faster steering. Did he like it?

"That depends. We change a lot of customers' cars from the faster steering to the slower. The fast can be too fast. I had to send one of my people to Italy to change a customer's car there. The guy was going 135 mph, hiccupped, and the Corvette changed lanes! That's scary stuff. It's way too fast for high-speed driving, unless you're a Formula One driver."

How about normal driving?

"The fast steering is okay. It's great for slalom. I don't mind it for everyday use. But you've got to stay on top of it. If you're driving down the freeway and you jiggle the wheel when you're looking around, you're changing lanes."

Guldstrand Engineering in Culver City, California, is where Dick hangs his business hat. It's not a big company, but it's still probably the Corvette aftermarket suspension center of the universe. Here's how Dick described his business.

"My business came directly from the knowledge gained in racing. The company has three divisions. There's an engineering group, a customer service group and a catalog group. The engineering group works on the development of products, not only for us, but for other companies like Monroe, General Motors and Goodyear. The customer service group works mainly with the customers who bring their cars to us. The catalog group, where ninety percent of our business comes from, takes care of selling customers the special parts we offer.

"At one time we sold some of our parts through other dealers, but we don't anymore. We prefer to deal with all of our customers directly even if that means limiting our growth. I think we have an impeccable relationship with our customers."

The interview with Dick Guldstrand was closed by asking why he has stuck with the Corvette, through thick and thin, for all these years.

"I wouldn't have changed places for anything," explained the racing-suspension master. "I loved being the underdog and I loved racing the all-American sports car. It was me against the Ford factory and the Porsche factory and all the rest. I had the satisfaction of running my own show. That was all me. I built the cars. I drug them to the races. I drove the cars. I had a rag-tag pit crew. It excited the press. We were known nationwide. There was a lot of ego involved. Maybe it wasn't the smartest thing to do. Maybe there'd have been bigger money going to Ford. But I didn't look at myself as that kind of driver. I was more of a technician. I was the team manager, the mechanic, the engineer and the driver. Usually, that's not tolerated in a racing program. Nor should it be. Each guy has to have his own job.

"The other thing was, I was so in love with the Corvette. Do you remember when the '63 came out? I mean, I thought I'd seen cars before. It was so beautiful! I could have kissed Bill Mitchell. I was so excited about that car. The *Corvette* is the car I wanted to be seen in."

4
Fast driving

Corvette. Fast driving. Tough to think of one without the other. Is that the affirmation of a genuinely great sports car? Or, the affirmation of a great Chevrolet sales job?

Both, of course. The marque has the heritage, the bloodline, the history. Belonging to the Corvette fraternity requires no more than ownership, and there are no driving-skill dues. Ownership alone makes the statement. That this is enough for many demonstrates how completely the Corvette has come to dominate American sports car culture.

Image alone isn't enough for some owners. They know their Corvettes have enormous performance capabilities. Some want to time-warp themselves back to the fifties, or sixties, and experience, or re-experience the sensations of that era. Some want to experience the awesome capabilities of newer Corvettes.

Fast driving.

There are Corvette owners who don't wish to compete in any fashion. They would, however, like a better hands-on understanding of how to drive their Corvette properly, how to control it in unexpected situations, and maybe how to comprehend better the terminology and descriptions filling every car magazine's road tests.

Books, videos and driving schools

How best to accomplish all this? Reading helps with the terminology, and some race drivers were self-taught with practice.

Video cassettes featuring the likes of Mario Andretti and Jackie Stewart are entertaining and educational. The Stewart offering, *Behind the Wheel with Jackie Stewart*, emphasizes Stewart's legendary smoothness, something all drivers should strive for.

Good as books, practice and cassettes are, nothing beats a driving school. No, not the one taught in high school by the industrial arts instructor. A driving school taught by professional race drivers.

Two such schools have gained excellent reputations. They are the Bob Bondurant School of High Performance Driving and the Jim Russell Racing Drivers' School.

Headquarters for the Bondurant School is at Sears Point International Raceway near Sonoma, California. The Russell school is headquartered in Salinas, California, but offers training at a variety of tracks in the United States, Canada and England. Although the majority of students who attend both schools have no professional racing aspirations, courses at the Russell school are oriented more toward the would-be professional. An indicator is

It doesn't take a bit of driving skill to be a Corvette owner. But it takes quite a lot of skill to drive a Corvette anywhere near its potential. Hands-on instruction from professionals in a controlled environment is the safest way to go fast quickly.

---

that Russell students train initially in single-seat Formula Fords, later in custom-built "Formula Russells." Russell graduates can go on to race in the Jim Russell Graduate Series or the Jim Russell Pro Series, professional races held in conjunction with major weekend racing events.

The Bondurant school utilizes single-seat Formula Fords for its several professional courses too, but it also has courses designed for the less ambitious. There is a one-day advanced street-driving course, and a two- or three-day high-performance driving course. These require students to bring their own cars or rent cars from Bondurant (Mustangs at present). One advantage of Bondurant's approach is that instructors can ride along with students and point out corrections, or take the wheel and let the students observe. With a student-instructor ratio of about three to one, the ride-along aspect is important.

The Russell approach is based on the theory that amateurs and professionals, at any level, learn best in real race cars. There's good logic in that, but I selected the Bondurant school, partly because of Bob Bondurant's reputation as a great Corvette racer, and the three-day high performance driving course, because a Corvette could be used. The Corvette was a base-suspension 1984 model with manual transmission. Let me tell you what it was like, starting with my arrival the first day.

Getting started at Bondurant's

In the waiting room, an old video tape of *To Tell The Truth* was on and, sure enough, there was Bob Bondurant along with two imposters trying to stump the panel as to the identity of the real racer and driving school president. One voted correctly, on the basis of Bob's comment about Paul Newman's Datsun sponsorship, but the other three blew it. Bondurant looked very professional. In the ensuing banter with Gary Moore, Bondurant commented that women made better students at his school than men because the ladies had fewer bad habits to break and less need to demonstrate their skills. But the men wound up going faster, according to Bondurant, because women reach a point where they just say, "Well, that's fast enough for me."

There were about twenty students in all and we gathered in a reception room for introduction of the instructors, all of whom were professional racers with varying degrees of experience. Then each student introduced himself and gave some personal background, including the intent of taking the course. The Grand Prix course people had racing on their minds, either with definite plans or to see if they realistically had the potential. The four other students in my course all professed to be there to learn better how to drive and appreciate high performance street machines. After everyone got acquainted, we received a tour of the track and facilities, then returned to the classrooms and divided into our class groups.

This first classroom stint was the longest of the three-day period, lasting over an hour. The school stresses driving time, and the total time spent in classes probably didn't amount to three hours for the full course. But the first class was important, because getting started correctly is a must.

The Corvette dominated the early discussions. The instructors, Parker Johnstone and Bill Mack, commented that American cars were designed back in pothole country, and the suspensions were so soft "they fell over on their door handles out here." But the Corvette was the exception. They'd had a chance to drive a new Corvette at Sears Point earlier, and it holds the unofficial record for production cars at 1:57 minutes for the two and one-half mile track. The next fastest were the Porsche 928 and Camaro Z-28, both at 2:02.

Driving position

Instruction started with the correct driving position. Bondurant stressed a bent-arm style, not locked elbows. When driving with arms straight, the logic goes, the back is pulled away from the seat, and the shoulders steer instead of the arms. This is less precise and increases fatigue in both shoulders and lower back.

The correct steering wheel grip was said to be at the nine and three o'clock positions, not the ten and two others, including Jackie Stewart, recommend. The grip itself should be firm enough to maintain control, but not a death grip.

Because of the new Corvette's quick steering, it is possible to leave the hands at the nine and three positions for a lot of cornering. If more steering input is needed, here's how it's done. For a left, the left hand is released from its nine o'clock position and slid over the wheel to just above the right hand. The right hand grip is released and the wheel is pulled a half turn to the left by the left hand. At this point, a half turn of steering has been

Top: The Bob Bondurant School of High Performance Driving recommends a driving position with elbows bent and hands at nine and three o'clock on the steering wheel.
Bottom: Photos of Bob Bondurant in action, here at Riverside in 1961, invariably show his determined yet relaxed driving style. Bob Bondurant gave driving instruction to actors for the movie Grand Prix, an experience that led him to eventually start his now-famous driving school. Dave Friedman photo.

---

dialed-in, and the hands are back at nine and three. If more steering is needed, the same action is repeated, and the wheel is unwound the same way in reverse. The beauty of the technique is that one hand is always gripping the wheel and the other is ready to. Maximum control! If the base position were anything but nine and three, more hand movement would be required. Less control.

The Corvette's steering wheel felt as if it was designed with Bondurant in mind, because the hands came naturally to the correct positions when resting them on the center supports. But the Bondurant instructors and

Bondurant's technique for steering emphasizes control. It works best for cars with fast steering, so the 1984-88 Corvettes are ideal. Older Corvettes require more steering input and hand changes.

Bondurant himself expressed preference for the traditional spoke placement that allows hooking the thumbs over the spokes with hands in the nine and three position instead of resting the bottom of the hands on the spokes as with the Corvette.

Applying the Bondurant thinking to other Corvette steering wheels, the 1953-55 wouldn't work well because the spokes are right at nine and three. Spokes in the 1956-62 models are a bit above ten and two positions, so they're perfect for hooking the thumbs over to leave hands at nine and three. The spokes of 1963-82 wheels were nearly back to nine and three for better instrument visibility, so hooking the thumbs puts the hands just below the proper position, but not by much.

Bondurant's folks preached smoothness constantly and made a strong point of shifting smoothly. A strong grip on the shifter and lightning shifts were taboo. An open hand technique where the shifter is shoved forward with the palm and pulled back with the fingers was demonstrated and recommended because soft grips lead to smooth shifting. Again, the Corvette's shift knob seemed perfect for the technique, but the instructors said they preferred something less severe because you "can't always grip it the way you're supposed to." Makes you realize how difficult the designer's job is.

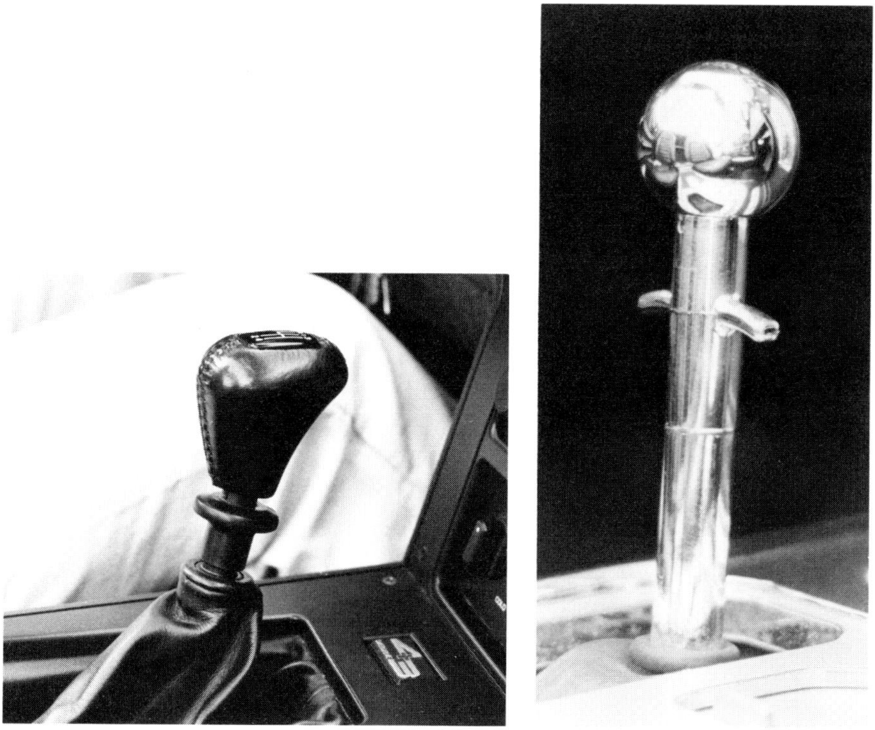

The shift knob shape of 1984-88 Corvettes fits the hand perfectly, but Bob Bondurant's instructors still preferred the symmetry of a round ball . . . exactly what nearly every previous Corvette has had.

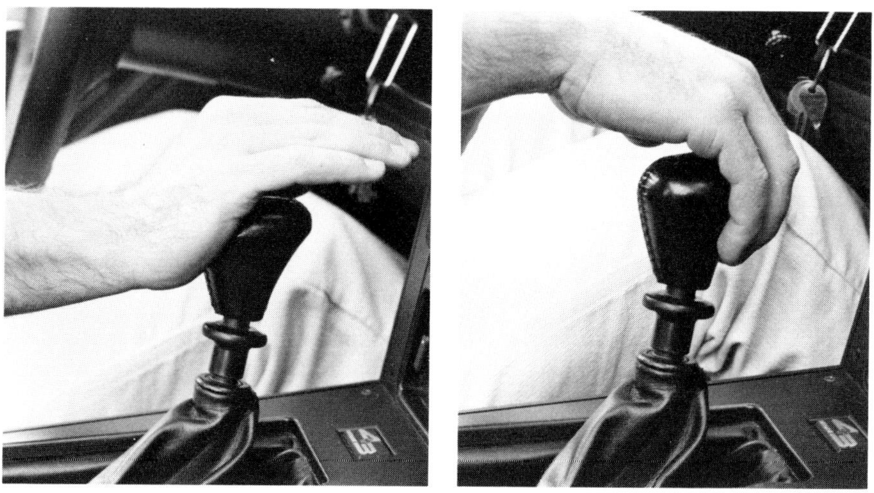

A tight grip on the shift knob encourages hard shifting and invites transmission problems. Avoid it by using the push-pull technique shown.

Heel-and-toe downshifting the 1984-88 Corvette works best by pivoting at the heel. The footwell is narrow, but the pedal arrangement puts the brake closer to the firewall than the clutch, permitting good heel-and-toe action.

Shift knobs in earlier Corvette models with manual transmissions have been round balls, just what Bondurant's instructors prefer.

Heel-and-toe downshifting

Heel-and-toe double-clutch downshifting is a must for race cars to preserve the transmissions. Here's how Bondurant's training manual described it:

"This is the technique of operating the brake and gas pedal simultaneously with the right foot while clutching with the left foot. If your car has an automatic transmission, this does not apply to your kind of driving.

"Heel-and-toe downshifting allows you to brake and match the engine rpm with rear wheel rpm, which allows for smooth downshifting.

"First you start to squeeze on the brake pressure to slow down the car. Then pivot the heel or side of your right foot onto the throttle, maintaining even brake pressure. Depress the clutch with your left foot, moving the shift lever into neutral on your way to the next gear. On releasing the clutch halfway, squeeze the throttle on to bring up the engine's rpm. Depress clutch again, quickly, and shift into the next lower gear. Release the clutch smoothly, then pivot your right heel off the throttle back to below brake pedal and continue trailing brake. This technique is called heel-and-toe downshifting with double clutching."

Understeer and oversteer

Next came a discussion of understeer and oversteer, which the instructors handled especially well. We all knew that understeer is when the front of the car breaks traction in a curve and starts to plow. Oversteer is when the rear breaks loose and starts to come around. A little oversteer doesn't bother an experienced driver because steering adjustment will control it, but it scares the bejesus out of a passenger who thinks a spinout is coming. An understeering car often isn't even evident to a passenger, but gives drivers white knuckles, because the driver feels the loss of steering control.

A common misconception about understeer and oversteer is that every car can be conveniently dropped into either category. Truth is, any car can be made to do both. It is true some cars have dominant tendencies one way or the other. Once a rear-engine Porsche starts to oversteer, it does so with a vengeance. Add to that the power of Porsche's turbo and its ability to break traction in the middle of an oversteering situation, and you've got a car whose controllability is beyond some drivers.

All manufacturers set their production cars up to understeer initially, no matter what inherent tendencies are present. The reason is safety, though that sounds like a contradiction. Understeer is quickly felt by any driver because steering control is affected. The nonprofessional reaction is always the same. The foot comes off the accelerator and heads for the brake. This is the proper action because weight is shifted forward to the front tires, the tire-to-road contact increases, and steering is restored. If ordinary cars oversteered initially, the average driver would still react the same, the weight would shift forward away from the rear which was already losing adhesion, and the car could spin.

Race drivers hate understeer. They want control and plowing front tires mean that steering control is all but gone. Oversteer is controllable. It can be corrected by steering the front with the steering wheel, or the rear with the accelerator.

The understeer and oversteer discussion was a prelude to one of the Bondurant course's fundamentals, that of properly getting around a corner. I got the conversation sidetracked to the relationship of engine placement to understeer and oversteer characteristics.

The instructors felt that front-engine, rear-drive cars were ideal for control by average drivers, and also great on the racetrack. These kinds of cars are especially fun to drive if, like the Corvette, they have ample power available to the rear wheels.

Front-engine, front-drive cars are fine for the general public because the weight bias makes them understeer and the unknowing driver will get out of trouble by the get-off-the-gas reaction. But for competition, they're less desirable. The power and steering are at the same end, and the rear just sort of follows along.

Rear-engine cars were deemed the least desirable for average drivers. No matter what tricks the manufacturer plays with the front geometry to give it initial understeer, when the rear eventually breaks loose, it'll come around in a hurry.

In racing, rear-engine cars can do well. They steer at one end and have power applications at the other, and the engine weight is over the driving wheels for great traction. But they require a skilled driver.

Mid-engine cars are wonderful for racing because they also have the engine weight near the tractive wheels, yet have even weight distribution that permits controlled understeer and oversteer. But when even a skilled driver loses a mid-engine car, it can spin violently out of control because of the central weight concentration. If a car has weight concentrated at one end, it's much easier to control a spin at 180 degrees or 360 degrees. The instructors' consensus was that a mid-engine car is the hardest to spin, but once it does, it's the hardest to control.

I was one who moaned when the Corvette didn't go mid-engine. But the Bondurant guys shed a new light on the subject. The Corvette's front-engine, rear-drive configuration is safer in average hands, true. But besides that, they got it around their own track faster than cars that aren't designed for anyone average. The low hoods of mid-engine cars can yield some neat shapes, but you've got to admit Chevy's logic was pretty solid.

Cornering

The class training got back on track as we learned the Bondurant technique for cornering. Bondurant said a turn is just a connection between straights, that it isn't so important how quickly you corner per se, but how fast you can go exiting the turn. If you exit just a mile per hour faster than the car alongside, that advantage will be carried through the entire straightaway. The idea, says Bondurant, is to sacrifice a little speed early in the turn so you can rocket out of it.

Turns vary, but in general this is accomplished not by taking the corner in the Interstate-style sweep, but by entering a little deeper than feels normal, then cutting back across under acceleration. It doesn't feel normal, because it's not the shortest way through. People tend to drive corners the way they'd run track, searching for the path with the least number of strides.

As the car cuts across the corner, there's an area where the car will be closest to the inside edge called the clipping point, or apex. At Bondurant's, one hears "apex" a thousand times. Judging where it is and hitting it properly in each turn is the trick to fast corner exits.

Let's talk through a turn and try to fit this all together. This description is for a sweeping 180 degree turn. As the turn is approached from a straightaway, first we need to get slowed down. The brake is squeezed on, quickly increasing to hard pressure but short of locking the wheels. If a downshift is necessary, it's done heel-and-toe during hard braking and before turning. The hard braking has shifted the weight bias forward to the front wheels for maximum steering input. As the turn is started, the foot comes off the brake, but not completely. Light pressure is maintained through about the first third of the turn. This is called trailbraking, and its purpose is to keep the weight bias forward, while the turn is being dialed-in.

At about a third into the corner, the trailbraking ceases, and acceleration begins. Two things flag this critical moment. The steering input is complete, and the rear is light, just on the verge of breaking loose. The throttle must be eased on, not stomped. Since the rear is on the verge of letting go, easing the throttle on transfers weight to the rear, and the tires bite in. A sudden surge of power to the rear would break the tires loose and start a slide.

All the while, the driver's eyes have been searching out the apex, about two-thirds through this turn. You aim for it and blast by it under acceleration. Done properly, the car should be completely stable at the apex. As you accelerate out of the turn, the steering wheel is gradually unwound, using as much track as possible.

With the instruction complete, we were ready to head for the track and try this apex business out for real. Sounded simple enough.

Bondurant had a section of the Sears Point track set up as a handling oval, a continuous loop with a hairpin at one end and a faster, increasing-radius turn at the other. We left our transmissions in second gear to better

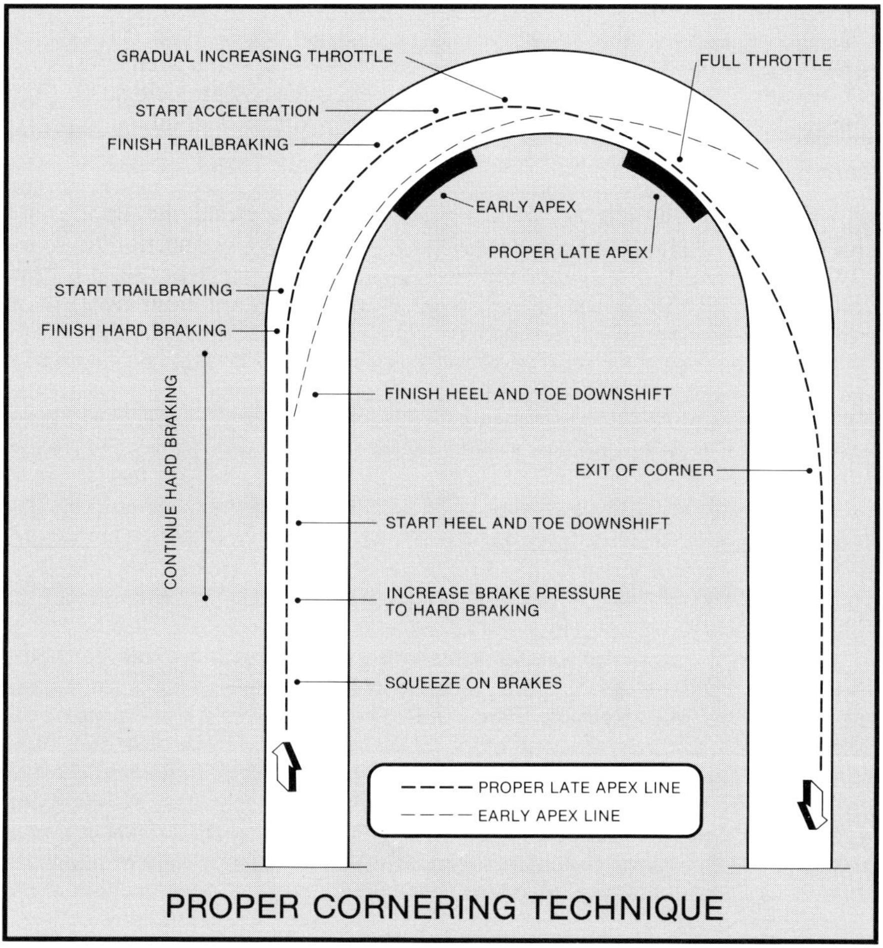

concentrate on the proper hand and seating positions, and the proper line through the corner. The braking points and apex areas were marked with cones, and we practiced on the handling oval for the balance of the day, which was for several hours.

As the instructors had warned, the natural tendency was to cut into the corner too soon (in effect an early apex), jab the throttle before the steering was set, then go through a series of throttle-steering-braking corrections. It was difficult to enter a turn as deeply as Bondurant stresses, because it felt unnatural and slow. But once it was mastered, the increase in exit speed was astonishing.

The Corvette was a delight to drive on the handling oval. The other cars required extra hand movement on the steering wheel to get through the hairpin, but the Corvette's quick steering didn't. I couldn't feel a trace of lean in the corners, confirmed later by my fellow students, but I could observe the other cars leaning as I followed them through. The Corvette was easy to control when off line. As I was learning the correct cornering techniques, I made mistakes, but the Corvette forgave them. If the rear hung out, a twitch of the steering wheel and a little throttle finesse brought it right back. The car was so forgiving it may lure some drivers into thinking they're better than they are.

The fast, precise steering of the 1984 Corvette would definitely have an advantage over earlier Corvettes with manual steering; they'd require much more steering effort and steering wheel turns.

We gave the cars quite a workout on the handling oval. The electronics were indicating six mile-per-gallon fuel economy with the Corvette. The front wheels of Howard Lasky's 944 were black with brake dust. Smoldering clutches and brakes stunk up the pristine California air. But the Corvette was clearly the quickest car on the oval and offered no complaints. I could tell it was happy.

On the second day, we spent a short time in the classroom reviewing the handling oval and the mistakes we were making, then went back to the oval to warm up. From there, we went to the far end of the Sears Point drag strip to practice heel-and-toe downshifting. Then it was back to the oval again.

Up until this time, everyone had only been using second gear in the oval, but then we started downshifting in every corner and using second and third. This threw a wrench into the smoothness we'd managed to develop and we had to start over somewhat. After a few hours, we were all sailing pretty smoothly.

The Bondurant instructors were superb. Parker Johnstone and Bill Mack were with us at all times. Two instructors for five students meant plenty of personal attention. They didn't take an overly critical posture. They allowed us to drive alone most of the time, make mistakes, sense the corrections, and make them. But once in a while, an instructor would wave a student into the pits, jump into the driver's seat for a few laps to point out a problem, then ride shotgun to see if the student had picked it up. It was a group class, but each student progressed at his own pace. The instructors maintained good balance of being there when needed, and keeping their distance otherwise.

Skid pad

In the afternoon of the second day, we got to play Jim Rockford on the skid pad. This is a 100×300-foot asphalt rectangle that's oiled up daily with a water-based goo. It was like a frozen supermarket parking lot without the shopping carts and light poles. Bondurant had a couple of Ford LTDs set up with special controls to enable the instructor to throw the car into a spin so the student could correct it. We learned how to control a slide at both 180 degrees and 360 degrees. There was also a simple exercise of locking the front wheels and turning the steering wheel lock to lock to demonstrate that with the brakes locked, there was no steering control. It just plowed straight ahead.

Accident simulator

Next was the accident simulator. This consisted of stoplights in three lanes. We approached the lights in the center lane at speeds from 30 mph to 45 mph, and at what seems like the last instant (seventy-seven feet away, actually) two lights went red and one went green. We were instructed to lift off the throttle when the lights changed and swerve into the green-light lane (the

Bondurant Driving School training involves hundreds of laps on a closed loop. The repetition is necessary to develop trailbraking technique and heel-and-toe downshifting, and to correctly judge each corner's apex.

---

reds signified blocked lanes), then ease the throttle back on when straightening out. Lifting the throttle shifted weight forward so the tires had some bite for the swerve, easing back on shifted weight back to the rear when the second correction was made. The Corvette's lightning quick steering made this a breeze.

Then the instructors threw us a curve. Sometimes all lights would go red and you had to make a brakes-locked panic stop. This complicated matters. However, the Corvette's brakes were terrific, so I figured there'd be no problem.

Big problem. Over 40 mph, neither I nor any of the others could get stopped before the lights. The best any of us could do was stop about halfway under. Some went a car length, or several, past.

As it turned out, it was virtually impossible to get stopped short of the lights, and that was the point of the accident simulator. In accident situa-

Corvettes are in the top ranks of sports cars for stopping power and it's been that way since 1965. An excellent new caliper design made its debut in 1984, then came anti-lock brakes in 1986. For the 1988 brakes shown, the calipers were redesigned again and a heavy-duty option, developed through showroom stock racing experience, was optional. Despite the Corvette's awesome ability to stop, Bondurant's training emphasized that many accidents can be steered around.

---

tions, drivers slam the brakes on, lock the wheels, lose steering and control, and crash. The Bondurant exercise is designed to illustrate that accidents can be steered around. If an accident looms, a driver should definitely get off the accelerator to transfer weight to the front wheels for better steering. This is instinctive. Then the driver must decide if there's a way to get around the problem before locking the wheels. If the choices are a ravine on the right and oncoming taffic on the left, it may well be that a panic stop is the only option. Locking the brakes can be the fastest way to stop in a straight line. But steering ability will be lost.

Note that this Corvette didn't have the antilock braking introduced in the 1986 model. The intent of antilock is to prevent the wheel lock and loss of steering just described.

Putting it together

The third day of training started with warmups on the oval to hone our coordination, and more skid pad work, then it was off to the "first loop" for what was as close to real racing as you'll get short of real racing.

The first loop is a 1.1-mile section of Sears Point and includes turns one through six. If you someday decide to take the high performance driving course at Bondurant's, you'll have the choice of two or three days. Take the three days, because the third is when you'll put what you've learned to use. Understanding how to bring out the abilities of a high performance road car, and your own, is best done in a repetitive track environment.

After a few hours of this, some real confidence was built up, but also a genuine respect for those who really do this well. A couple of laps with Bondurant at the wheel of the '84 Corvette and myself in the passenger seat emphasized that.

The first loop experience also reinforced the three needs stressed by the instructors: smoothness, consistency, and concentration.

Smoothness feels slow at first, but it's much, much faster. There's never been a great racer who drove in a herky-jerky manner.

Consistency is more obvious. Once you've dialed in a corner, and it could take hundreds of laps to do it correctly, it has to be driven consistently to run consistently fast.

Most important is concentration. Lapses in concentration can be devastating. Just a little mind drift can put you into a corner incorrectly. If you know what right feels like, you'll recognize wrong. Steering wheel death grip sets in. With skill (maybe luck for beginners) and a correction, you're on your way. But one lapse can lead to a series of errors and lead to big trouble. As students, we had the luxury of pulling off the track in a designated area to rest a moment, to get our minds cleared, and to start fresh. The pros must concentrate for hours.

At the end of the third day, we had the opportunity to follow our instructors on a few laps of the full 2.5-mile Sears Point course. The advanced students were on it (we shared the short loop with them but had the right of way), and it was evident some of them were getting very quick. Then we piled into our instructors' cars for a couple of "hot" laps. All this was meant to prime our pumps to return for advanced training.

The students in my class all had a marvelous time. We developed a confidence in each other as we tooled around on each other's heels. We trusted each other and switched cars so that we could sample and compare. The Corvette took a back seat to nothing, especially in that environment.

The great thing about the Bondurant course, its professionalism aside, is the opportunity to train in your own automobile. You can bring your Corvette, any model year, and take the entire three-day course with it. One caution. You drive your own car at your own pace, but it is difficult not to push hard. If you're concerned about three days of hard use, you have the option of training in Bondurant rentals for the first two days, then switching to your Corvette for the third, on-track day.

Taking a driving course in your Corvette will give you and it a sense of unity you wouldn't have thought possible. Don't do this in a Corvette you don't plan to keep.

5

Amateur and professional racing

Years from now, auto historians will look back on the Corvette's first half-century and conclude there were two golden ages of Corvette racing.

The first started in 1956. Chevrolet decided to give its floundering Corvette a performance image and to use racing as the means. That officially came to an end in 1957 when GM sponsored an Automobile Manufacturers Association (AMA) racing ban. As we all know today, Chevrolet continued to covertly encourage outsiders to race its products with support that ran hot and cold.

No one can agree when the Corvette's first glorious racing period came to an end. By 1957, Corvettes were dominating their SCCA B-production class to the extent that these became races between Corvettes, not between Corvettes and competing marques. Still, the Corvette events were love 'em or hate 'em crowd favorites, usually held last on the day's schedule so spectators wouldn't leave early.

The arrival of the Cobra late in 1962 ended the Corvette's domination, despite its all-new 1963 chassis. Still, the Cobra was fierce, yet fragile, and Corvettes continued winning races and championships into the mid-sixties. So some might end the Corvette's first golden racing age at 1962, while others would extend it a few more years. But all agreed. It ended in the sixties.

The first golden age was important for the performance image it handed the Corvette, but also for the real improvements in production cars, resulting directly from racing experience.

These two big factors marking the Corvette's first golden racing age—the image of an unbeatable racer, and direct refinement of the production cars based on racing experience—also mark the Corvette's second golden racing age, which started in 1985, when the Corvette became the dominating marque in the SCCA Showroom Stock series, both in endurance Pro Showroom Stock, and in SCCA's traditional Showroom Stock. The latter was a direct descendant of the regional SCCA races dominated by Corvettes a quarter-century earlier.

What's more, the interest in Corvettes with first-golden-racing-age pedigrees is undergoing a tremendous resurgence, compliments of a booming phenomenom called vintage racing.

This all bodes well for the Corvette enthusiast interested in racing either as spectator or participant. At the entry level are events sponsored by

The new chassis introduced in 1984 brought the Corvette back to road racing glory in the SCCA Showroom Stock series. Here, Baker Racing Team leads the pack in 1986 at Mosport, Canada. Rich Chenet photo.

Is Corvette vintage racing popular? It's the biggest thing to hit the Corvette hobby in a decade. This is the Laguna Seca track near Monterey, and the annual historic auto races held there are the biggest anywhere. In 1987, Chevrolet was the featured marque.

local, unaffiliated Corvette clubs, events perhaps no more ambitious than a little friendly Sunday slalom on an empty parking lot. The National Council of Corvette Clubs (NCCC), the parent organization for some 400 Corvette clubs in the United States, takes this a step further by sanctioning local events, including drag racing and timed road course runs, then sponsoring regional and national championships. The National Council of Western States is not affiliated with the NCCC, but provides similar activities for clubs in the western United States.

All-Corvette events suit many owners just fine. But let's spend the remainder of this chapter discussing the two big racing activities already mentioned in which the Corvette is only one marque of many, albeit a most important one: vintage racing and SCCA Showroom Stock racing.

Vintage racing

Vintage racing is for older cars and the sport itself is nothing new. The Europeans have been at it for decades, and some groups in the United States have held little gatherings with their beloved old racecars over the past thirty years. But the current vintage mania traces its roots to 1974 when Steve Earle launched his first annual Monterey Historical Races at Laguna Seca, near Monterey in California. When asked if there was a personal motivation behind promoting an event like Monterey, Mr. Earle (no relation to Harley Earl) replied this way:

"The motivating force behind the idea was that I wanted to see cars restored and not rotting away. There were racecars either in somebody's

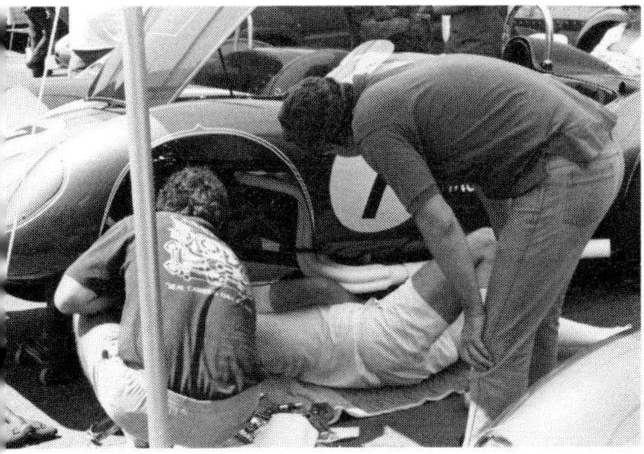

Anyone who loves racing would drool over the automobiles that show up for vintage race events. The 1987 historical races at Laguna Seca welcomed antique vehicles along with two of the original five Grand Sport Corvettes, a couple of Jim Hall's Corvette-powered Chaparrals, Corvette-powered Scarabs, even a Corvette-powered Cheetah.

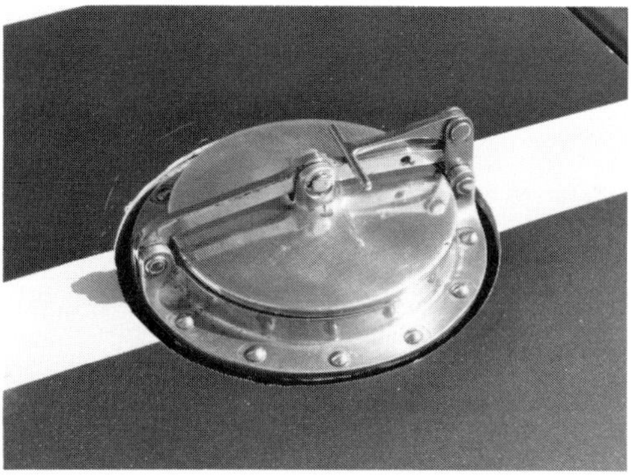

Steve Earle, who can be given much credit for starting the vintage racing boom with his first Monterey event in 1974, decided early-on that the goal of these events should be to share wonderful old race cars with enthusiasts. His open-pit policy permits spectators to view and photograph the race cars up close. This is an experience no Corvette enthusiast should miss.

backyard getting rained on, or in a museum—deteriorating in either case. We felt if there was an opportunity for guys to run them, they'd restore some of these things.

"That's been our underlying philosophy all along in putting on these events and doing this. The idea of Monterey's being a spectator event was that there were an awful lot of people who couldn't afford to own the cars. Even though many racecars have been very cheap at times, most people still can't afford to have and restore one. It's always been that way. We hoped that the people who were fortunate enough to have an old racecar would like to share it. So at Monterey, enthusiasts get to see cars and get to see them run.

"It was a mutual getting together of enthusiasts to share these great old racecars. That too, has been an underlying philosophy for us through the years. As the sport has grown, people see it now more as real competition. It certainly is competition, but we want the underlying motivation to be, 'Hey, if you're lucky enough to own a Grand Prix Mercedes-Benz, you're one of less than a handful; it's great that you'd bring it out and let the rest of us see it run and enjoy it and have a good time with it.' "

The Monterey Historical Race has the reputation of being the premier vintage event, but it is by no means the only one. In fact, the number of vintage races seems to double each year. The SCCA has a vintage race calendar, and it's becoming fashionable to include a vintage race on the schedules of all sorts of professional events.

Vintage racing offers fantastic spectator opportunities for Corvette enthusiasts to view, hear and experience the kinds of Corvettes only previously seen in pictures. But is it serious racing? Some drivers take it seriously; most don't. Tracks, for the most part, don't want serious racing and strongly discourage it (Monterey gives the trophy to the *last* place finisher). After all, the owners and drivers of vintage Corvettes tend not to be hot-shot racers on the way up, but more the middle-aged businessman interested in the comraderie, the ambiance, of days past. And some of the cars are extremely valuable, reaching into the millions of dollars.

What's required for a Corvette enthusiast to enter vintage racing actively? Does the Corvette have to have a bona fide racing history? Do you have to have some kind of racing license?

So far, few event sponsors require a check that the vehicles have real racing credentials, though this will probably tighten in the future. As for the drivers, it depends on the sanctioning organization or event promoters. Some could care less who or what raced. Others, like Steve Earle's operation at Monterey, reserve the right to turn applicants away.

"We look at everything," explains Steve Earle. "We really prefer, and now request, that guys have gone through one of the major driving schools, and taken the full course, not a one-day deal. On top of that, we prefer also that they've participated in vintage races somewhere. We recognize that when you do feature marques, you're going to have a couple of guys who used to race but haven't in a long time. If they've never done any driving before, we request that they attend one of the driving schools before we'll consider them. For everyone's safety, we want to avoid the guy who's never been there before. Not that they're going to go fast and lose control of their car. What happens is that they haven't picked up the habit of looking in their mirrors or of being aware of traffic. They can very easily pull over at a point when somebody's pulling along beside them, and crunch."

So there it is, a new twist to the Corvette hobby. Vintage racing offers a terrific way for Corvette enthusiasts to relive the Corvette's first golden age of racing.

SCCA Showroom Stock racing

If the present is more appealing than the past—not that anyone can't be involved in both—consider SCCA Showroom Stock. First a little background.

The SCCA sponsors two types of showroom stock racing, the two types being administered by separate groups within the SCCA. Think of these two as professional and amateur (or club), though the term *amateur* is something of a misnomer.

The words "showroom stock" alone usually refer to the amateur SCCA races. Classes are GT, A, B and C. Corvettes race in GT, the fastest class. The amateur races are short sprint-type, perhaps twenty laps, and nearly always lasting less than an hour. The cars are street-driveable with just a few modifications for safety. A roll bar is required, but the rules are very specific concerning how it's mounted so that it doesn't increase structural rigidity. Goodyear was the sole supplier of tires for Corvettes in 1984, but since then other manufacturers like Michelin and B.F. Goodrich offer tires for Corvettes, both in performance-street versions, and in softer-compound types for competition.

Entrants are permitted to service their engines, but may not make any changes for the sake of improved performance. Engine blueprinting is prohibited.

Other cars in the Corvette's amateur Showroom Stock GT class include IROC Camaros, Trans Am Firebirds, Mustang GTs, Nissan 300ZX Turbos and Porsche 944 Turbos, the latter being the Corvette's strongest competition.

Regional races are run all over the country, with a national runoff each October at Road Atlanta to determine champions in each class. Corvettes run very strongly in amateur SCCA Showroom Stock racing, but aren't invincible because of the short nature of the races. Corvettes did dominate in 1985 and 1986, but Porsche wound up winning the runoffs at the end of 1986, and pulled about even in wins in 1987.

It's a different story for the professional SCCA series, Pro Showroom Stock. These are endurance races from four to twenty-four hours. The series started in 1985 and the Corvette was never defeated in any race during the first two seasons. Other cars ran with the Corvettes, but none could last, because of the Corvette's superior durability. More on this later.

In order to encourage aftermarket suppliers to participate in the Pro series, some modifications are allowed. Any wheel can be used, so long as it's the same diameter and offset as the factory originals. Seats, tires and shock absorbers can be changed.

When it started in 1985, the Pro series had the same GT, A, B and C classes as amateur Showroom Stock. Also, the GT classes of both were restricted to cars with a base price under $30,000. Because the Corvette dominated so completely in 1985, the SCCA changed the classes for 1986. The slowest class, C, was combined with B. The Corvette and the Porsche 944 Turbo, the only real competitors in the old GT class, were bumped up

Top: This is John Heinricy piloting the Morrison-Cook entry at Riverside in 1985. Rich Chenet photo.
Center: Powell Motorsport Advanced Driving School fielded this striking entry, shown here in the pits at Mosport, Canada, in 1986. Drivers were R. Ceppos, R. Spenard, R. Fellows and M. Engelage. Rich Chenet photo.
Bottom: Bobby Carradine drifts a Morrison-Cook Corvette through a turn at Portland in 1987. The Honda CRX, though no match for the Corvettes, did dominate its own class. Rich Chenet photo.

Top: This was Mid-Ohio in the wet with the R. A. Motorsports Corvette entry leading a Porsche 944. The 944s were Corvette's stiffest competition in the Showroom Stock series. Rich Chenet photo.
Center: Bill Cooper and Mark Dismore handled driving for this D. R. Motorsports Corvette, here at Portland in 1987. Rich Chenet photo.
Bottom: This dramatic wheel-off-track photo shows the Powell Motorsport Advanced Driving School Corvette at Sears Point in 1987. Rich Chenet photo.

Fierce and not-too-stock, this Corvette was built for the IMSA 1987 GT racing series.

You didn't have to go IMSA racing in 1987 to own a Corvette with awesome power. The beautifully detailed Callaway Twin Turbo Corvette delivered a whopping 345 hp, 0-60 mph times of 4.6 seconds and a top speed of 175 mph in 1987 trim. It carried the standard Corvette warranty and could be ordered from select Chevrolet dealers. It was RPO-B2K and added $19,995 to the window sticker.

Who says Corvettes aren't tough? This RCG Racing Corvette soldiered on at Mid-Ohio in 1986 despite some . . . er, damage. Rich Chenet photo.

---

to a new super class, SS. The $30,000 cap was dropped for the SS class with the hopes of attracting cars like the Ferrari 308 and 328, the Lotus Esprit Turbo, and the Porsche 928S4. But except for the 944 Turbos, no one showed. Corvettes won all the races again. Perhaps the other marques, all with proud racing heritages, didn't care to have their street cars compared to Corvettes in this brutal and visible showcase.

There's a reason Corvettes have been so successful in amateur and professional SCCA racing beyond the Corvette's superb chassis design. Just as in 1956, Corvette engineers have eliminated weaknesses by incorporating racing's lessons right into production vehicles. This isn't sales hype, it's real.

Corvette engineers learned in the Corvette's first showroom stock endurance race that it wasn't bulletproof. Wheel studs broke, hubs cracked and ignition modules shorted under extreme heat conditions. Chevrolet contracted Morrison & Cook Racing for testing.

Listen to Doug Robinson, who coordinates the engineering side of the Corvette's racing programs.

"We had Morrison & Cook simulate six-hour races. When we got to the point where we could run six hours without a failure, we went on and on until we got to twenty-four hours. We tested until we got a failed part. We'd take that part back to the lab to analyze it. While we were testing, we instrumented the cars and measured all the loads, temperatures and operating conditions. Then it was back to the lab where we simulated those loads and found they were outside the realm of anything we saw in our proving grounds tests when we did the basic car development.

"So we set up lab tests, simulated the failures, determined the causes, made design changes to eliminate the failures and ran more tests to verify the failures were eliminated. By the end of the 1986 season, we'd fixed anything that could routinely fail in a twenty-four hour race."

To the delight of Corvette enthusiasts, all these improvements went right into production Corvettes. Which recent production Corvette model has benefited the most from racing? Here's Doug Robinson again:

"The 1986 was the first year where things from showroom racing began to benefit the production car. That year probably saw the most. In 1987, we added the power steering cooler to the Z51 purely because of racing. On the street, you'd never get it hot enough to cause a problem. But for racing, it was the last weak link. For 1988, we optimized the suspension geometry, and the new brake design was second to none compared to anything anyone's ever put on a street car."

Corvettes are involved in racing other than the SCCA Showroom Stock. For instance, the Corvette fielded a strong entry in the GTO class (over three liters engine displacement) of International Motor Sports Association's (IMSA) GT series in 1987. But of most interest to the majority of Corvette enthusiasts are those types of racing which most directly influence the production cars available to us all, so that's been highlighted here.

Amateur or professional, spectator or participant, Corvette enthusiasts can enjoy racing activities like never before.

Make that like once before.

6

Special concerns, problems, solutions and products

The joys of Corvette ownership are not without a price. There are special concerns and problems every Corvette owner confronts. Not all apply to all Corvettes; not all are exclusive to Corvettes.

Fuel

The owners of older automobiles must deal with the removal of lead from fuel at the direction of the United States Environmental Protection Agency (EPA). The amounts of lead added to motor fuels has declined in several steps with a goal of complete elimination. The lead phaseout was based on valid health considerations. However, Corvettes through and including the 1970 models, were designed to operate with leaded fuels. Owners of these models must compensate in some way for the subsequent changes in fuel composition.

Why was lead, tetraethyl lead to be specific, added to motor fuel to begin with? Two reasons. First, lead was an economical means of boosting octane. Refiners can achieve adequate octane without adding lead, but it takes more crude oil per gallon of fuel produced to do it. Using fuel with inadequate octane results in engine "knocking," regardless of the model year. Second, lead provided a lubricating "cushion" between an engine's exhaust valves and their seats to minimize wear. Use of unleaded fuel in pre-1971 Corvettes results in premature exhaust valve erosion.

Speaking to the valve wear issue, how much lead is required? It varies with the type of engine, engine temperatures, the type of use (high rpm require more lead), even the altitude (high altitudes require less). As a general guide, it's thought that 0.2 grams of lead per gallon of fuel will provide adequate valve protection for perhaps ninety percent of the pre-1971 vehicles in use. By 1987, only 0.1 gram of lead per gallon was permitted in leaded fuels, and that varied from station to station. By contrast, leaded fuels in the pre-1970 era had as much as 3.0 grams per gallon, or thirty times the 1987 maximum.

Lead in automobile gasoline helped both octane and valve wear, but dealing with the elimination of lead might involve a multiple approach. The choices fall into the categories of fuel mixing, additives and engine modifications.

Top: Lead substitute fuel additives contain compounds that mimic lead's valve lubricating properties. Without an additive, use of low- or no-lead gasolines can damage valves and valve seats in pre-1971 Corvettes.
Bottom: The octane levels of most pump gasolines aren't adequate for Corvette engines with compression ratios over 9.0:1. Octane boosters not containing lead lift octane but don't lubricate valves. Octane boosters with real tetraethyl lead both increase octane and lubricate valves, but future availability of additive products with real lead is uncertain.

During the lead phaseout period, some owners of older Corvettes found that mixing three parts unleaded premium with one part leaded regular yielded a fuel with adequate octane. Since the leaded was diluted with unleaded, valve protection was questionable. Mixing fuels continues to be a good idea so long as leaded fuels are available in any form, including aviation gas. Even if leaded fuel is unavailable, unleaded premiums may

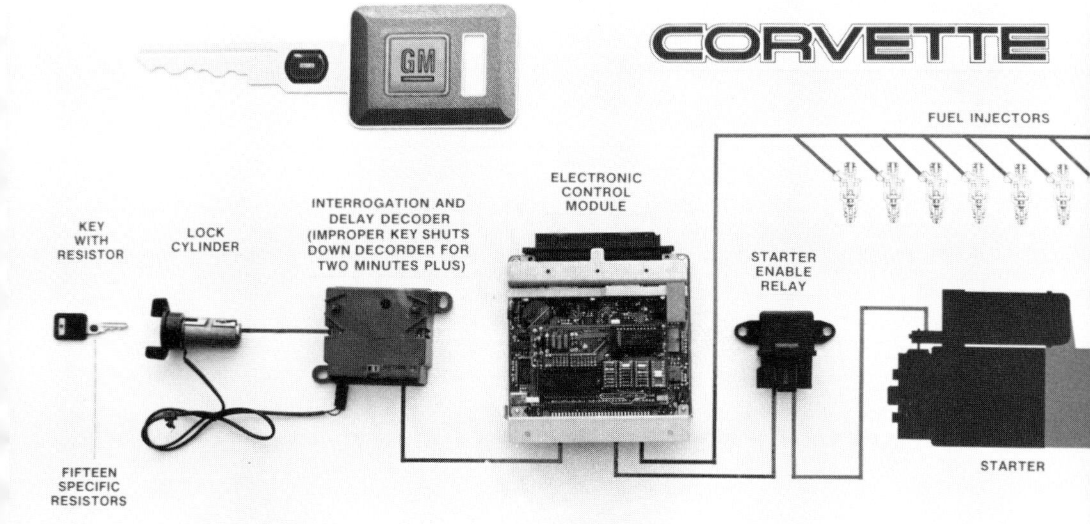

Top left: Owner-added alarm system features external key activation, but does require drilling body panels.
Top right: The Ungo Box windshield decal says this Corvette has a sophisticated aftermarket alarm system with such features as a remote arm-disarm transmitter, motion-shock detector, panic alarm and automatic rearm. A pager is optional.
Bottom: Chevrolet introduced VATS (vehicle anti-theft system) on 1986 Corvettes. The Highway Loss Data Institute says Corvette thefts were off 45 percent in the calendar year 1987, and it credited VATS for the decline.

---

satisfy octane requirements if combined with retarded engine timing.

Retarding engine timing does reduce engine performance and fuel economy, as do other options for reducing an engine's octane appetite. Octane requirement is largely dictated by an engine's compression ratio. Lowering the compression ratio can be done by changing pistons, or using thicker or multiple-head gaskets. Beware. Many rebuilt Corvette engines have had the opposite treatment. They've had heads or blocks milled, or thinner head gaskets were inadvertently used, which resulted in higher compression ratios.

Owners of older Corvettes who modify their engines for lower octane usage, should aim for about 9.0:1 compression. In your calculations, remember that compression ratio is nothing more than the volume differential created in the combustion chamber by the piston moving between its two extreme positions. Don't be influenced by the compression ratios of newer automobiles which are creeping back up (to 9.5:1 in the 1988 Corvette), because sophisticated electronic knock-sensors, in combination with electronic ignitions, are used.

The cylinder heads of earlier Corvettes can be modified to be compatible with unleaded fuels to eliminate the valve-burning problem. The common approach is to machine the valve seat and install a stainless steel insert. Valves are replaced with special Stellite high-temperature types. The cost for all this, excluding labor, is about $25 per valve. Exhaust valves are subject to the most wear, but owners who go this far might wish to do the intakes while they're at it.

The Lubrizol Corporation of Wickliffe, Ohio, the largest manufacturer of fuel and motor oil additives, has developed an additive it calls PowerShield, a sodium-based compound which mimics lead's valve-lubricating properties (it does nothing for octane). Lubrizol doesn't market its products directly to consumers, but scores of other companies buy PowerShield in bulk and sell valve-lubricating fuel additives containing it. No major oil companies currently blend PowerShield into pump fuels, but it could happen in the future if sufficient demand develops.

Real Lead, an additive offered by the Octane Boost Corporation, contains genuine tetraethyl lead. It protects valves and valve seats and boosts octane slightly. Because it contains the real McCoy, its future availability isn't assured. Nonlead octane boosters normally employ compounds of either manganese or aniline for raising octane. Without something else added, they don't protect valves.

Insurance

Fuel concerns affect owners of 1970 and older Corvettes, but insurance is a subject that affects virtually every Corvette owner. We all know insurance companies don't like to insure Corvettes. But why? Alan Blay, an independent agent and Corvette collector who's written about Corvette insurance in *The Corvette Restorer* magazine for over a decade, put it this way:

"Insurance companies would rather not insure a Corvette for liability due to the fact that many people can and do drive them hard and recklessly, thus increasing the accident potential. They would rather not insure a Corvette for collision because they are more costly to fix, in addition to the fact that the car is a symbol of pride and few Corvette owners will accept a cheap or shoddy repair. They would rather not insure a Corvette for comprehensive due to its ease of theft and high theft potential. In addition, the high performance drive train will fit right into another Chevrolet."

There are no cookbook formulas for obtaining Corvette insurance, and comments made here about premium costs have to be rather general because of the wide range dictated by area, age and driving record of the insured, and Corvette model. Virtually all auto insurance companies insure Corvettes, but not by choice. Some are required by their states to accept unwanted clients in a sort of lottery called "assigned risk." Others do so as a courtesy for existing customers, especially ones who have homes, businesses and other cars already insured. Soliciting Corvette business by insurance companies is virtually unheard of, and few will write new-customer policies for Corvette owners.

Insuring newer Corvettes, say under fifteen years old, is one thing. Insuring older models is another. Both are difficult, for similar and different reasons. Newer Corvettes tend to be driven and, as Alan Blay pointed out, experience higher accident and theft rates, and repair costs. Insurance companies can determine the value of newer Corvettes easily enough, and the frequency and cost of claims is documented to the hilt. Even at what appear to be astronomical premiums, insurance companies don't want this business.

For two of the same reasons—high value and high repair costs—they don't want older Corvettes either. There's a bigger problem with the oldies. Data on older Corvettes just doesn't distill neatly into an insurance agent's manual. Some are driven daily, some are trailered once a year to a concours. Some are outside in the rain, others are in safe-as-a-bank-vault storage. Some are valuable and some are priceless.

A handful of companies specialize in collector car insurance, though even they don't welcome Corvettes. And these collector policies are restrictive, often not permitting any kind of daily use. The best advice is to work through your personal insurance agent. If yours is an older model, remove the mystery and protect your investment by seeking a stated-value policy whereby you and your insurance company agree in writing what your Corvette is worth. With or without a stated-value policy, provide an appraisal, photos and documentation of components and condition.

For newer Corvettes, working with your agent is also the best advice. States with assigned-risk statutes make insurance available to anyone, but the cost is high. Conventional insurance with a well-established company is best if you can get it, and you probably can if you're an existing customer with a clean record. One way to soften the rates is to accept large deductibles for collision and comprehensive coverage.

If you don't already own a Corvette, you really must explore the insurance question before you purchase. The availability and cost could influence your choice of model year as much as the hardware.

Theft

Theft is related to the insurance issue, and is of special concern to Corvette owners. The many positive driving and owning characteristics detailed in this text explain in part why Corvettes rank so high on thieves' most-wanted lists. Another part of the theft problem is that, until recently, Corvettes have been too easy to steal.

The first Corvettes, the 1953-55 models, had no provisions for even locking the doors! There were no external door handles. Entry was gained by reaching through the side curtains and pulling a knob inside the door. Of course, convertibles are easy prey for thieves anyway.

Chevrolet made an alarm system optional for Corvettes in 1968, which became standard equipment in 1972. The system was improved for the 1984-85 models; yet all these, which activated the vehicle's horn, were no match for a competent thief. In 1986, a much-improved Vehicle Anti-Theft System (VATS) was introduced for Corvettes. This one used a special ignition key with an embedded pellet. Ignition contacts measured the pellet's electrical resistance and, if correct (there were fifteen variations), the Corvette would start. If wrong, the Corvette wouldn't start, and at least three minutes has to pass before another attempt could be made.

The time-delay element in the VATS system is why it's effective. It isn't foolproof, nor is any other antitheft device. The best one can do is delay the

Top: Be careful what you use to cover a Corvette in storage. High quality covers, like this one from Mid America Corvette Supplies, provide safe indoor storage because they don't form an airtight barrier. Mid America photo.
Bottom: After owning this 11,000-mile 1960 Corvette for twenty-five years, the original owner sold it in disgust after the paint was ruined by outside storage under an airtight tarp.

thief. Professional car thieves get their work done quickly, or they move on. There are any number of antitheft devices a Corvette owner can employ. There are alarms and pagers, defeats to prevent the engine from starting or running long, or mechanical devices to prevent pedal or steering wheel operation. They all work, but they're all beatable. The time delay of multiple systems is most effective.

Storage

This text is aimed at owners who drive their Corvettes. Still, there are times when the most ardent Corvette driver must put his Corvette away in storage. Given a choice, it's better to drive a Corvette, even infrequently, than store it without any attention for extended periods. But if it has to be done, here's how.

First, perform all routine maintenance like chassis lubrication, oil and filter change, and inspection of fluid levels. The radiator should be drained and filled with a clean antifreeze, antirust solution. For extra protection, use distilled water.

Some pros advice draining the fuel tank; others say fill it. The problem with draining is that the tank's inner surface could rust. The problem with filling is that over long storage, automobile gas can turn to a gel and leave a varnish accumulation throughout the entire fuel system. However, aviation fuel does not go sour, nor does "white" gas. And additives, like Sta-bil and Fuel Fresh, are said to prevent auto gas from spoiling.

Storage on "blocks" is easier on the tires and suspension and a good idea for duration longer than a few months. The old, original tires that came with Corvettes before Department of Transportation (DOT) markings were required are very valuable. They can be protected in storage by filling them with nitrogen gas.

The biggest danger of long-term Corvette storage is piston-ring seizure to the cylinder walls. Minimize the chances for this by removing the plugs and squirting oil directly into the plug holes. Turn the engine over with the plugs out to coat the rings and cylinder walls. A quicker way to accomplish this is to stall the car out by pouring oil into the carburetor. Better yet, do both.

It's a good idea to put boxes of mice killer in the engine compartment and interior to stop these critters from making your Corvette their new home. Dehydrant packets like those found with electronic gear should be spread around the interior if humidity levels are high in your area. Some owners put mothballs on cloth seats, but there will be an odor to contend with later.

The battery should at least be disconnected, and is best removed entirely. Convertible tops should be left up. Put the windows up, but leave a one-inch gap at the top. Cover the Corvette, but be absolutely certain the cover is one that breathes. Old cotton sheets are fine.

Removing a Corvette from storage is mostly a matter of cleaning and inspecting. Before moving it from its storage, look underneath for tell-tale leaks. Before trying to start the engine, even if the cylinders were coated with oil, squirt light motor oil, even penetrating oil, into the plug holes again to free a possibly frozen ring. Let it soak hours or even overnight. Put in fresh plugs or clean the old ones. Start the engine, bring it up to temperature, then shut it off and change the oil.

Leaks

Mention leaks to a Corvette owner and he'll think either of fluid leaks from the disc brake systems of 1965-82 models, or water leaks into the cabins of 1953-55 models, or 1956-62 models, or . . .

The disc brake leaks occurred around the seals of the caliper pistons. The 1965-82 Corvette disc brake system (a delete-disc option was available in 1965 only) used four pistons at each wheel, so each car had sixteen pistons in all. The piston material was anodized aluminum, the calipers were either nodular cast iron (early) or gray cast iron (late). If the anodizing came off the pistons, electrolytic reactions took place between the iron and

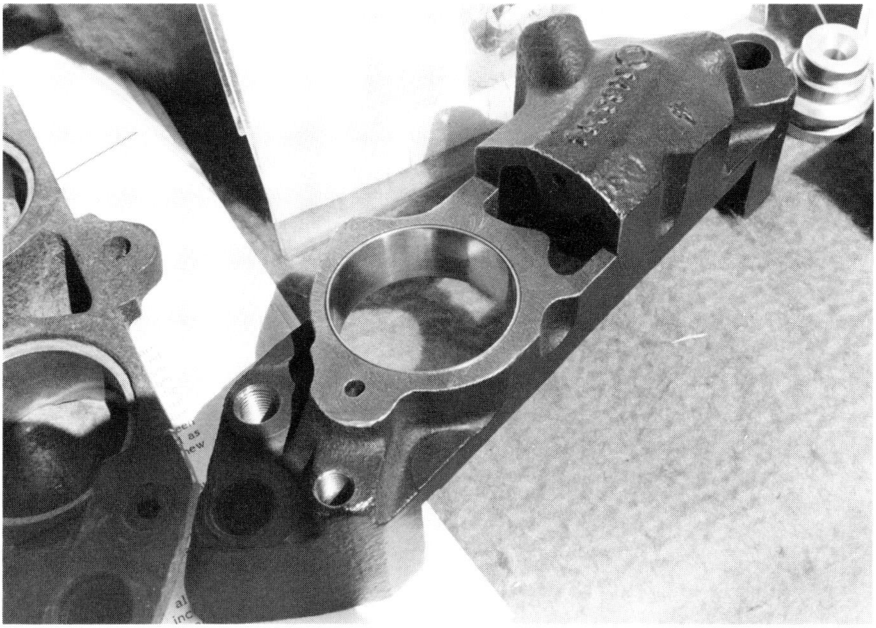

The disc brakes of 1965-82 Corvettes had great stopping power, but were prone to develop fluid leaks around the caliper bores. Entrepreneurs in the Corvette aftermarket solved this annoying problem by developing a process of lining the caliper bores with stainless steel sleeves. High quality manufacturers like Dr. Vette (example shown here), Stainless Steel Brake Corporation, Vette Products and others offer sleeved Corvette brake calipers with lifetime guarantees.

---

the aluminum. The bigger problem was that the iron rusted. As pads wore, the pistons moved toward the pads, taking new positions in the bores which had been exposed to the elements. As the seals came across corroded areas, they stopped sealing.

This was a terribly frustrating situation for Corvette enthusiasts, having a brake system that worked so well between very expensive visits to the repair shop. But the Corvette aftermarket solved it by sleeving the caliper bores with stainless steel, and substituting stainless steel pistons for the aluminum ones. Several very reputable companies offer this service, now expanded to include stainless steel emergency brake kits.

Water leaks into Corvette interiors are not so easily solved. The first Corvettes were the worst, incredible leakers, but the new models are virtually watertight. The problem decreased in an almost straight-line plot through the Corvette's life span. The 1956-62 models leaked, but not nearly like the first ones. Employees of the St. Louis Corvette plant have said that no 1963-67 Corvette ever completely passed the plant's high-pressure water test. Still, from 1963-on, leaks into the interiors of Corvettes were of the drip-drip, rather than the open-faucet, variety.

If an older Corvette hasn't been restored, its rubber seals and calked joints may have deteriorated. If it has been restored, it may have been done improperly. One example that comes to mind is a friend's 1965 which leaked

Corvette fiberglass bodies don't rust but their metal frames do. The area to check first is the rear of the side frame rails around the kick-up for the rear wheelwell. The cross support member forward of the fuel tank is the second most likely rust area. Sometimes localized frame rust can be repaired. Severely rusted frames have to be replaced entirely.

---

badly onto the carpeting from behind the instrument panel. Turned out, a new windshield hadn't been sealed properly.

The best thing an owner can do is make sure all rubber seals and calked joints are in tip-top shape, then aim between the clouds.

Rust

Rust? Haven't all Corvettes had fiberglass bodies? Yes, but they've all had metal in their chassis. Starting with the 1984 model, Chevrolet took great care to protect Corvette chassis metal subject to corrosion by special coatings. To date, the newer models show no evidence of serious chassis corrosion problems. But the conventional-style frames of all Corvettes from 1953 through 1982 were made of rust-prone steel. The main sections were boxed, ideal for trapping moisture. A petroleum-based paint, not especially long lasting or corrosion retarding, was applied only to the exterior frame

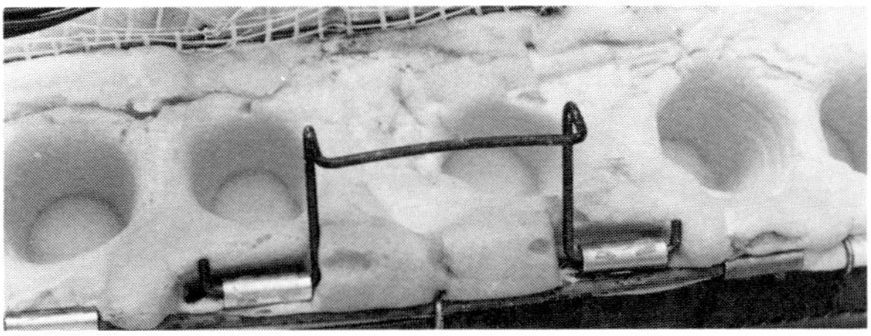

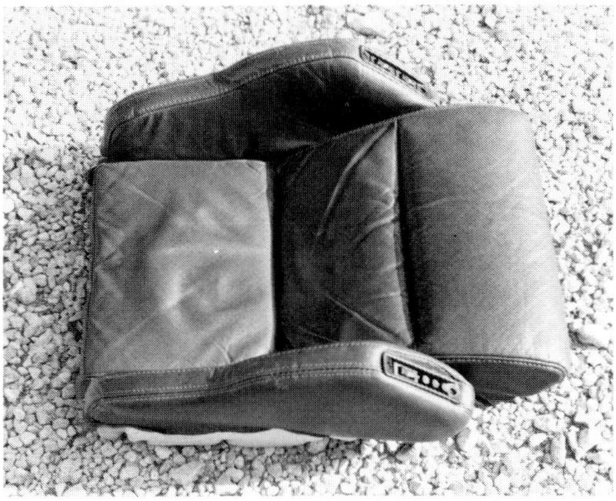

After 8,000 miles of careful use, this 1985 leather sport seat (driver side) shows considerable wear. A single spring clip at the front releases the seat cushion and bolster assembly. After simple pneumatic and electrical disconnects, the driver and passenger sides can be interchanged so that at least wear is evenly distributed.

surfaces, mainly for appearance. Rust through from the inside is very possible, and even frames that look good on the surface could be on the verge of rusting through from the inside.

A good body-off restoration includes cleaning the frame by either sandblasting or acid submersion. The disadvantage of sandblasting is that it is impossible to reach all the frame's inner surfaces. The acid bath reaches everything, but if not properly neutralized, new coatings don't stick well. Some restorers combine both techniques for cleaning. Then, a good rust-preventative coating is applied to all frame surfaces. If a natural factory appearance is desired, the original type coating is applied *over* the rust inhibitor.

If a Corvette frame has already rusted enough to require repair, there are a couple of options. Often the rust is localized (the most common area is the side rail just as it kicks up for the rear wheel housing), and a repair is possible. A section from a good frame can be welded in, or a repair section, recently available from aftermarket companies, can be used. Sometimes a frame's rust is so extensive that the entire frame has to be replaced.

Specialty products

A peek into any major Corvette aftermarket supplier's catalog will reveal any number of specialty products for other perceived Corvette short-

text continues on page 105

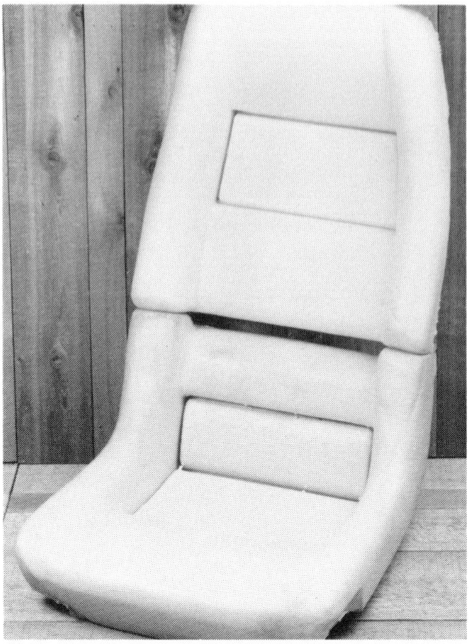

Left: Weather boots are form-fitting, ABS plastic footwells that protect carpeting from water, dirt, snow and salt. Eckler's Corvette Parts has them for 1968-88 models. Eckler's photo.
Right: What's the point of replacing seat upholstery if the foam under it is shot? Eckler's Corvette Parts offers replacement seat foam for 1958-82 Corvette seats. Eckler's photo.

Eckler's Corvette Parts sells hundreds of different Corvette items, but it's most famous for its fiberglass body panel manufacturing expertise. In addition to factory replacement panels, Eckler's pioneered one-piece panels like the 1975-79 front end shown to eliminate seam finishing. Eckler's photo.

Everyone's heard of sheepskin seat covers. Compared to leather or vinyl, they're cooler in summer and warmer in winter. These shown are actually called inserts because they don't cover the bolsters of 1984-88 seats. Eckler's Corvette Parts has full covers for these models too, and for 1968-82 Corvettes. Eckler's photo.

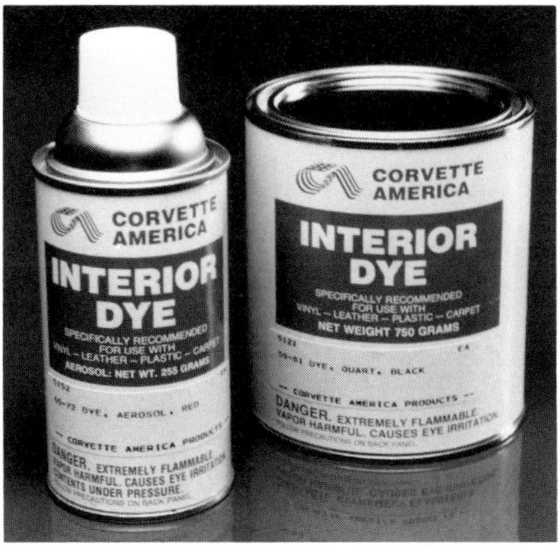

Left: Corvette America formulates, color matches, mixes and packages its own interior dye in quart cans or aerosol bombs. This flexible coating is intended for seat covers, plastic parts, door panels and dash pads. Corvette America photo. Right: For the finishing touch, Corvette America has stainless steel exhaust tips for all 1963-88 Corvettes. The 1963-73 tips replace factory tips with the same appearance. The later models didn't have bright tips from the factory, so the Corvette America tips add a new look. Corvette America photo.

Chevrolet offered genuine knock-off wheels for 1963-66 Corvettes (experts aren't convinced any 1963 was actually delivered with them) and a "bolt on" style for 1967 models. Chevrolet obviously didn't make enough, because the demand for originals is tremendous. Corvette America manufactures new versions of the wheels and, along with its dealers, is the exclusive source. Corvette America photo.

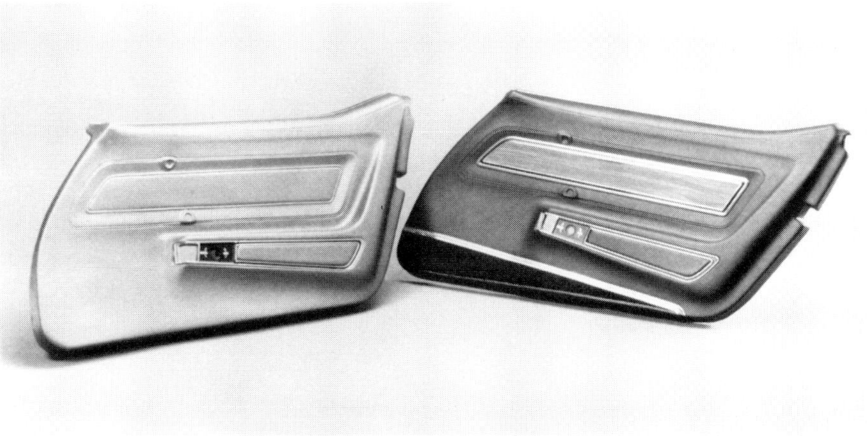

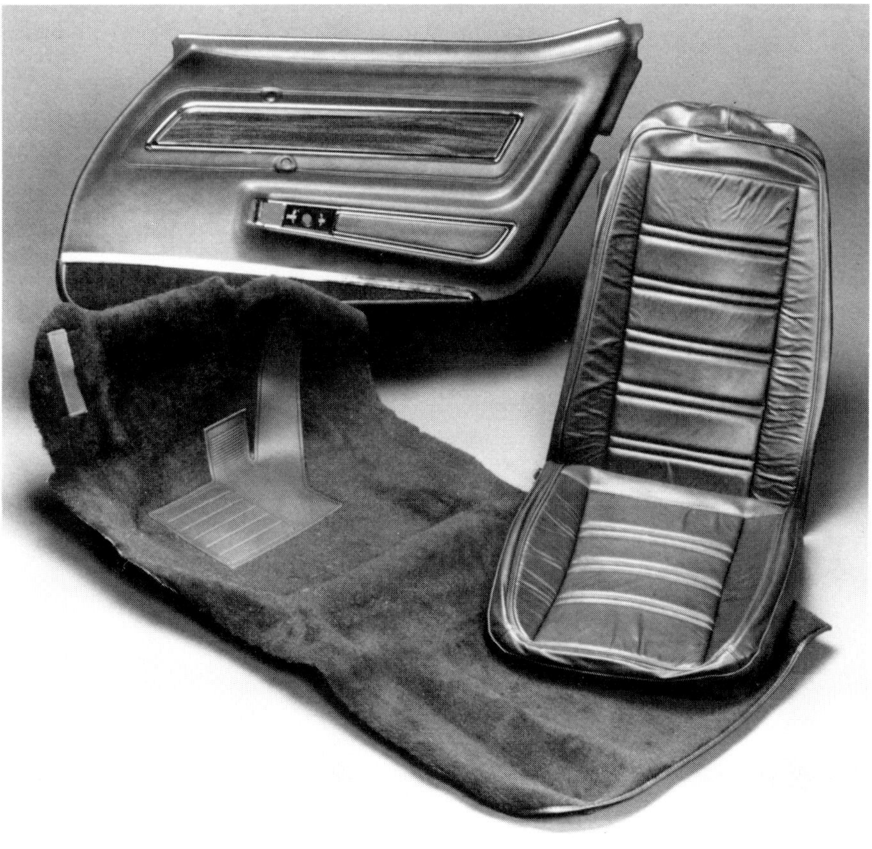

Top: Door panels were a weak link in Corvette interiors, especially the molded foam style that first appeared in 1965. Corvette America sells complete replacement door panels for most 1959-77 Corvettes, and less expensive "skins" that cover the old panels for 1963-76 models. Corvette America photo.
Bottom: If you need it all, you can save by purchasing interior sets consisting of carpeting, door panels and seat covers. Corvette America photo.

Here are some fun items from Mid America Corvette Supplies. How about Corvette leather shoes that'll even leave a Corvette imprint in the sand the next time you visit Acapulco? How about draw-string Corvette North Pole Anoraks? Better yet, how about a black Corvette Teddy with slip top and snap crotch? Mid America photos.

How have you gotten along without a genuine Corvette shower curtain, or without Corvette bar lights and clocks handmade of brass? Mid America photos.

No Corvette should be without its owner's manual, and Mid America Corvette Supplies has reproductions for 1958-1980 models. For those who wish an every-nut-and-bolt breakdown of their Corvette's body assembly, Chevrolet-authorized reprints of the assembly manuals the factory used are available from Mid America Corvette Supplies for 1956-57 and 1960-82 models. Mid America photos.

For a Corvette (1986 shown) with a little Ferrari personality, American Custom Industries offers this special side and rear treatment.

Top and bottom left: Mirror and front end masks provide protection from stone chips and bugs. Some owners swear by protective covers, but others won't use them for fear of paint abrasion from wind flap, or paint clouding if moisture is trapped.
Right: There's no controversy about reflective screens. They bounce sunlight out to protect the dash and keep the interior cool. The manufacturer of this one is The Canvas Works of Mountainview, California.

comings. Some of these are frivolous, others very ingenious.

The 1984-88 Corvettes have a very high sill to step over when entering. The sill is plastic, painted to match the interior color. It scuffs easily. At least three varieties of protector plates are available, all of them better than Chevrolet's solution of carpeting the sill, starting in 1988. Replacement sills are available for other models, but their need is debatable.

The narrow footwells of some Corvette models invite excessive carpet wear along the sides. ABS-plastic footwells are available to protect the carpeting, especially in rain and slush conditions. Styles for 1968-88 models can be purchased, but the 1984-88 cars need them most. A variation of this idea is a cargo liner for the luggage area of 1984-88 models.

Putting anything on a Corvette's finish that doesn't breathe, like front end masks, is asking for trouble. However, if one's Corvette was subjected to gravel roads or unusually nasty bugs, occasional use of a mask would be warranted. Don't leave it on any longer than necessary.

If exposure to sun is worrisome in an owner's area, the aftermarketeers have many solutions. There are dash covers, seat covers, window covers and window shades.

Beyond all this, there are scores of doodads that don't fix anything, but might add to a particular model's convenience, like dash pouches and elbow cushions. There are appearance items like real wood inlays for instrument panels, and replacements for the 1984-88's "breadloaf," the protrusion on the passenger side dash area which looks as if it housed an air bag but didn't.

Thanks to the Corvette's popularity, there's almost no end to the flow of aftermarket products to fix and enchance these great cars. It's a shame that concerns like fuel, insurance, and theft have to butt in on all the fun. Coping with unleaded fuel is an irritation we can live with because the benefits in air quality are worth it. Insurance cost and insurance availability, and theft result in part because of the very qualities that make Corvettes so desirable.

Ironic.

7

Events, organizations and assistance

An advantage of the Corvette's popularity is the sheer number of events and organizations, entirely separate from the Chevrolet Motor Division, that cater to enthusiasts.

Shows and clubs

Hardly a weekend passes without a Corvette show taking place somewhere. These range from small gatherings of a local club to huge events drawing over 10,000 enthusiasts. No other marque offers anything even closely resembling this phenomenon.

The shows trace their origins to another almost-unique facet of the Corvette hobby, Corvette clubs. Sure, other marques have clubs. But none offer the numbers and the diversity found in the Corvette movement. There are at least 700 individual Corvette clubs around the United States, more in Canada, and even a sprinkling in foreign countries. It all started in the late fifties and was encouraged by *Corvette News*, the periodical published by Chevrolet for Corvette enthusiasts since 1957.

Shows began when clubs held little events for driving or showing their Corvettes. Some incorporated swap meets. By the early seventies, it was an idea whose time had come. Some club shows grew into major events, as did other independent shows with no club affiliation. The big shows normally are built around a concours, parts vendors and individuals with everything imaginable for sale, and sometimes an all-Corvette auction.

There are scores of well-run, smaller shows, each with a flavor and attraction of its own. But by the late eighties, the show scene, so far as majors were concerned, had boiled down to the big three. In March, it was Knoxville, Tennessee, held nearly all indoors and featuring a gigantic Corvette auction. In June, it was Bloomington, Illinois, the world's premier Corvette event. In August, it was Corvettes at Carlisle, Pennsylvania, known for its tremendous swap meet.

Check with enthusiasts in your area to see which events in your area offer what you seek. If you want to see it all in one event, go to Bloomington. Bloomington's concours is superb because of its emphasis on factory originality, and the quality of its judges. Virtually all Corvette models can be viewed in as-built condition. The swap meet is first rate, as is the Corvette auction. A Special Collection of rare Corvettes is displayed for two evenings in a beautiful hall at the Illinois State University, about three miles

Big Corvette shows, like this one held each June at the Bloomington, Illinois, fairgrounds, feature an auction, a swap meet and a great concours to draw all models of Corvettes for judging. Bloomington has additional attractions, like a "special collection" display of unique Corvettes, and workshops for learning Corvette restoration techniques.

NCCC activities appeal to action-minded Corvette enthusiasts. The 1987 national convention held at Columbus, Ohio, offered drag races at National Trail Dragway and timed road race runs at Mid-Ohio racetrack.

---

from the fairgrounds where the other activities are held. Instruction workshops for specialized Corvette repairs and restoration techniques can be attended with advance reservations.

The NCRS and NCCC

If an owner is interested mostly in 1953-72 model Corvettes, and factory originality is important, membership in the National Corvette Restorers Society (NCRS) is a must. Started in 1974 by seven owners of early Corvettes frustrated by the lack of accurate restoration data, NCRS membership now exceeds 7,000. It also sponsors shows throughout the year, including an annual convention at a different location each year. Small swap meets are a part of some NCRS shows, but remaining true to its charter, NCRS events are built around members' Corvettes.

Like Bloomington, NCRS judging doesn't compare Corvettes to each other, but against a factory-original standard. Some of the same people are both Bloomington and NCRS judges. The quality of their knowledge is best exemplified by the increase in market value a Bloomington Gold or NCRS Top Flight award conveys.

Another advantage of NCRS membership is its quarterly magazine, *The Corvette Restorer*. As club magazines go, Corvette or otherwise, it is

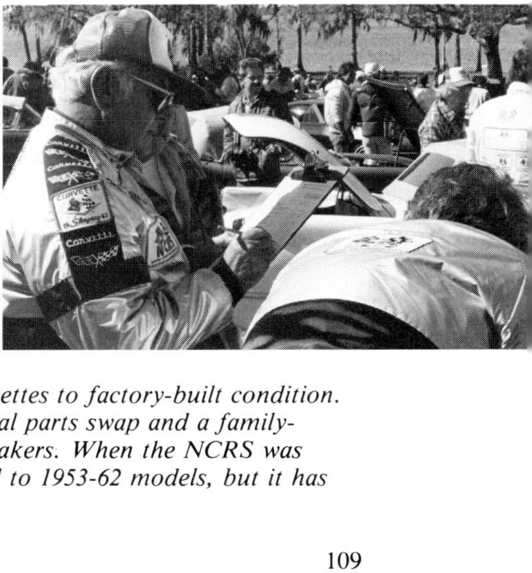

NCRS was founded to preserve and restore Corvettes to factory-built condition. NCRS meets typically feature a concours, a casual parts swap and a family-oriented awards dinner with interesting guest speakers. When the NCRS was founded in 1974, its range of interest was limited to 1953-62 models, but it has since expanded to include through 1972s.

There was a time, not too many years ago, when Corvettes like these sat around used car lots waiting for someone to pay very meager prices to take them home. Now some owners consider them even too valuable to drive.

---

simply the best in both presentation and content. The magazine alone is worth twice NCRS annual dues, currently $21.00 in the United States.

With membership around 10,000, the largest organization for Corvettes is the National Council of Corvette Clubs (NCCC). This is the parent organization for the majority of Corvette clubs. NCCC is less restrictive in its appeal to Corvette owners than Bloomington or the NCRS. NCCC welcomes them all. Where a Corvette with nonoriginal components is doomed in a Bloomington or NCRS event, NCCC has concours classes to accommodate any Corvette. Originality is used to determine classes, but once classed, Corvettes are judged purely on condition and cleanliness.

The NCCC also emphasizes driving events, both at the local levels and at its national convention held annually in different locations. These include drag racing, slalom, rally and even timed runs on road race courses.

Chevrolet dealers and Corvette shops

Speaking of Corvette dealers, another benefit of the Corvette's popularity is the number of independent Corvette shops. Some come and go, but others are in for the long haul and offer excellent full service to Corvette customers. Check with other enthusiasts in your area to find the good ones. And don't ignore your local Chevrolet dealers. Many recognize the Corvette's appeal and have specially trained Corvette technicians, and sometimes even special Corvette parts and supply setups. Chevrolet dealerships do tend to be most interested and knowledgeable about newer models, while the independent Corvette shops often cater to owners of earlier models.

Don't look for a lot of assistance directly from Chevrolet with respect to older models, either. Chevrolet, like all auto companies, does maintain a public relations staff. But they are most helpful to automotive media people, not to individual Corvette owners. Understandably, the people at Chevrolet are most enthusiastic about the automobiles they sell today and tomorrow, not those sold twenty years ago.

Publications

One of the most frequent types of questions Chevrolet hears from Corvette owners centers on the number of models produced with some specific combination of options. How many red 1965 fuel-injected convertibles were sold with such and such options? Because the possible combinations are in the hundreds of thousands, Chevrolet, or anyone else, can't answer. But publications available to enthusiasts can help with these and other types of data. Good monthly magazines offer a wide selection of advertising from companies and individuals for Corvettes and Corvette parts. A list of the books and magazines every Corvette enthusiast will find useful is provided in the rear of this book.

In summary, the Corvette's popularity leads to more popularity. Where other marques get relegated to museums or junkyards, Corvettes live on. More and more, old Corvettes are being restored, driven, enjoyed. Even organizations like NCRS, which caters mainly to restoration-minded owners, emphasize driving. NCRS's top Corvette recognition, the Duntov Award, requires a mechanical test to ensure a Corvette functions as well as it looks. New Corvettes already challenge and win against the world's best sports cars.

The Corvette's appeal for driving and owning is clear. Corvette owners, intentionally or otherwise, become part of the marque's history, it's controversy, its pains, its pleasures. The shows, the organizations, the clubs and the literature are all frosting on the cake.

Ed Cole saw something special in the first Corvette prototype way back in 1952. Yet, in his wildest imagination, he couldn't have forseen where his gut-feel request to "get" the Corvette for Chevrolet would lead.

When all is said and done though, what matters in the end is the car. Does it deliver?

It does.

1953-55

1953

Base price
1953: $3,498.00
1954: $2,774.00
1955: $2,774.00 (6-cylinder)

Production
1953: 300
1954: 3,640
1955: 700

Brakes
Drum, front and rear.

Transmission
Two-speed automatic; three-speed manual (1955 only).

Body style
Roadster, soft top only.

Suspension
Front: Coil springs, tube-type shock absorbers, stabilizer.
Rear: Leaf springs, tube-type shock absorbers, solid axle.

Dimensions
Length: 167.3"
Width: 69.8"
Height: 51.5" (over windshield)

Instruments
Speedometer
Tachometer (with cumulative RPM counter)
Odometer
Oil pressure
Water temperature
Fuel
Ammeter
Clock

Historical overview
• 1953 production of 300 was the lowest in the Corvette's history.
• 1953 models were available only with white exteriors and red interiors.
• The third Corvette built, serial number E53F001003, exists and is privately owned, but the first and second have never surfaced and are believed to have been destroyed by Chevrolet in testing.
• Horsepower was increased during 1954 from 150 to 155.
• The V-8 engine was introduced to Corvettes in 1955.
• All 1953-54 models, and 1955 models with six-cylinder engines, had six-volt electrical systems.
• All six-cylinder 1955 Corvettes had automatic transmissions.

Positive driver/owner factors
• Wide, comfortable seats.
• Conventional trunk.
• Wide footwells.
• Excellent pedal placements.
• Cumulative RPM counter on tachometer.
• Excellent fresh-air ventilation.
• Low octane fuel requirement.

Negative driver/owner factors
• No seat adjustments except driver fore-aft.
• Long, non-adjustable steering column.
• Poor tachometer and secondary instrument locations.
• Windshield distortion at corners.
• Limited rear visibility.
• Limited leg room.
• Dated suspension.
• No door locks or exterior door handles.
• Severe water leaks into cabin.
• Six-volt electrical systems (except V-8).
• Recirculating-style heater.
• Unwieldy soft top mechanism.
• Plastic side curtains.
• Difficult entry (windshield post)
• Leaded fuel required.

1956-57

1957

Base price
1956: $2,900.00
1957: $3,176.32

Production
1956: 3,467
1957: 6,339

Brakes
Drum, front and rear.

Transmission
Two-speed automatic; three-speed manual; four-speed manual (1957 only).

Body style
Roadster; soft top and/or removable hardtop.

Suspension
Front: Coil springs, tube-type shock absorbers, stabilizer.
Rear: Leaf springs, tube-type shock absorbers, solid-axle.

Dimensions
Length: 168.0"
Width: 70.5"
Height: 51.0" (over hardtop)

Instruments
Speedometer
Tachometer (with cumulative RPM counter)
Odometer
Oil Pressure
Water Temperature
Fuel
Ammeter
Clock

Historical overview
- The instrument panel and most chassis components carried over from 1955, but otherwise the 1956 was all-new. New items included roll-up glass windows with optional power assists and external door handles and locks.
- A four-speed manual transmission was first optional in 1957 Corvettes.
- Fuel injection was first optional in 1957.
- A removable hardtop was first optional in 1956 Corvettes.
- All 1956 engines were 265-cid V-8s with horsepower ratings from 210-hp to 240-hp.
- All 1957 engines were 283-cid V-8s with horsepower ratings from 220-hp to 283-hp.
- Visually, 1956 and 1957 models were almost identical. One difference was the rearview mirror. The 1956 adjusted with a thumbscrew; the 1957 required a wrench.
- Fore-aft seat travel adjustment was added to the passenger seat starting in 1956. It had previously been on the driver seat only.

Positive driver/owner factors
- Potent engines.
- Conventional trunk.
- Wide footwells.
- Excellent pedal placements.
- Cumulative RPM counter on tachometer.
- Fresh air heaters (except early 1956).
- Excellent fresh-air ventilation.
- Simple manual soft top operation.
- Ample legroom.

Negative driver/owner factors
- Narrow, flat seats.
- No seat adjustments except fore-aft.
- Long, non-adjustable steering column.
- Poor tachometer and secondary instrument locations (except 1957's RPO-579E).
- Windshield distortion at corners.
- Dated suspension.
- Moderate water leaks into cabin.
- No dead pedal.
- No reverse lockout with 4-speed.
- Limited headroom.
- Leaded fuel required.

1958-60

1960

Base price
1958: $3,591.00
1959: $3,875.00
1960: $3,872.00

Production
1958: 9,168
1959: 9,670
1960: 10,261

Brakes
Drum, front and rear

Transmission
Two-speed automatic; three-speed manual; four-speed manual.

Body style
Roadster; soft top and/or removable hardtop.

Suspension
Front: Coil springs, tube-type shock absorbers, stabilizer.
Rear: Leaf springs, tube-type shock absorbers, radius rods (1959-60), stabilizer (1960), solid axle.

Dimensions
Length: 177.2"
Width: 72.8"
Height: 51.0" (over hardtop)

Instruments
Speedometer
Tachometer
Odometer
Oil pressure
Water temperature
Fuel
Ammeter
Clock

Historical overview
- The 1958 model was the first Corvette with four headlights.
- The 1958 Corvette had chrome strips on its trunk and simulated louvers on its hood, but both were removed the following year.
- A reverse lockout was added to four-speed transmissions in 1959.
- Seat belts were factory-installed in 1958-60 (dealer-installed previously).
- The 1958-60 interior, and exterior body panels were redesigned.
- The instrument cluster in 1958-60 Corvettes put all instruments (except the clock) directly in front of the driver.
- The bumpers of 1958-60 Corvettes anchored indirectly to the frame to offer better protection compared to previous models.
- Steering columns were shortened in 1958.
- A reverse lockout was added to four-speed manual transmissions in 1959.
- Rear radius rods were added in 1959.
- A rear stabilizer was added in 1960.

Positive driver/owner factors
- Potent engines.
- Conventional trunk.
- Wide footwells.
- Excellent pedal placements.
- Excellent instrument locations.
- Excellent fresh-air ventilation.
- Simple manual top operation.
- Ample legroom.

Negative driver/owner factors
- Narrow, flat seats.
- No seat adjustment except fore-aft.
- Non-adjustable steering column.
- Small tachometer face.
- Windshield distortion at corners.
- Dated suspension.
- No dead pedal.
- No reverse lockout with four-speed(1958).
- Limited headroom
- Leaded fuel required.

1961-62

1962

Base price
1961: $3,934.00
1962: $4,038.00

Production
1961: 10,939
1962: 14,531

Brakes
Drum, front and rear

Transmission
Two-speed automatic; three-speed manual; four-speed manual.

Body style
Roadster; soft top and/or removable hardtop.

Suspension
Front: Coil springs, tube-type shock absorbers, stabilizer.
Rear: Leaf springs, tube-type shock absorbers, stabilizer, radius rods, solid axle.

Dimensions
Length: 177.7"
Width: 70.4"
Height: 51.5" (over hardtop)

Instruments
Speedometer
Tachometer
Odometer
Oil Pressure
Water Temperature
Fuel
Ammeter
Clock

Historical overview
- The 1961-62 Corvettes were the last Corvettes with conventional trunks.
- The 1961-62 Corvettes were the last with solid rear axles.
- The 1961 Corvette was the last to offer two-tone paint until 1978.
- 283-cid engines were last available in Corvettes in 1961. 1962 was the first year for the 327-cid.
- Wide whitewalls were last used on Corvettes in 1961.
- Four-taillight treatment, common to all Corvettes since, was first used in 1961.
- The transmission tunnel was reduced in width by 20% in 1961-62 models.
- The large-bar grill treatment, common to 1953-60 models, was changed to mesh for 1961-62 Corvettes.
- The 1962 Corvette was the last model with exposed headlamps.
- 1962 was the last year for an optional power soft top.
- Aluminum radiators were used in all 1961-62 Corvettes.
- The automatic transmission case in 1962 Corvettes was aluminum, a first.

Positive driver/owner factors
- Potent engines
- Conventional trunk.
- Wide footwells.
- Excellent pedal placements.
- Excellent instrument locations.
- Ample legroom.
- Excellent fresh-air ventilation.
- Simple manual top operation.

Negative Driver/Owner Factors
- Narrow, flat seats.
- No seat adjustment except fore-aft.
- Non-adjustable steering column.
- Small tachometer face.
- Windshield distortion at corners.
- Dated suspension (but improved).
- No dead pedal.
- Limited headroom.
- Leaded fuel required.

1963-64

1963

Base price
1963: $4,037.00 (convertible)
1964: $4,037.00 (convertible)

Production
1963: 21,513
1964: 22,229

Brakes
Drum, front and rear

Transmission
Two-speed automatic; three-speed manual; four-speed manual.

Body style
Coupe; convertible with soft top and/or removable hardtop.

Suspension
Front: Coil springs, tube-type shock absorbers, stabilizer, unequal length A-arms.
Rear: Transverse leaf spring, trailing arms, lateral struts, tube-type shock absorbers, stabilizer.

Dimensions
Length: 175.1"
Width: 69.6"
Height: 49.8" (coupe)

Instruments
Speedometer
Tachometer
Odometer
Trip odometer
Oil pressure
Water temperature
Fuel
Ammeter
Clock

Historical overview
• The 1963 model introduced a completely new interior, new exterior and new independent rear suspension.
• 1963 was the first year for a choice of coupe or convertible body styles.
• Hidden headlights first appeared on Corvettes in 1963.
• Air conditioning, power brakes and power steering were first optional in 1963.
• Leather seating was first optional in 1963.
• Knock-off aluminum wheels were first offered as options for the 1963 model, but it's doubtful any were actually factory-installed until 1964 due to leaks.
• AM-FM radios were first optional in 1963. Early 1963 had AM signal-seeking.

Positive driver/owner factors
• Potent engines.
• Adjustable seat backs (1963).
• Good pedal placements.
• Superb instrument locations.
• Superb instrument legibility.
• Adjustable steering column (underhood).
• Simple manual top operation.
• Excellent ride quality.
• Air conditioning available.
• Ample headroom.
• Vent windows standard.
• Generous luggage space (coupe).

Negative driver/owner factors
• No seat adjustment except fore-aft (1964).
• Poor cabin ventilation (coupe).
• Steering column not adjustable from inside.
• No dead pedal.
• Limited rear visibility (coupe).
• Difficult luggage access.
• Limited luggage space (convertible).
• Difficult battery access (some engines).
• Difficult spare tire access.
• Narrow footwells.
• Minor interior water leaks.
• Leaded fuel required.

1965-67

1966

Base price
1965: $4,106.00 (convertible)
1966: $4,084.00 (convertible)
1967: $4,240.75 (convertible)

Production
1965: 23,562
1966: 27,720
1967: 22,940

Brakes
Disc, front and rear (except delete disc option in 1965 only).

Transmission
Two-speed automatic; three-speed manual; four-speed manual.

Body style
Coupe; convertible with soft top and/or removable hardtop.

Suspension
Front: Coil springs, tube-type shock absorbers, stabilizer, unequal length A-arms.
Rear: Transverse leaf spring, trailing arms, lateral struts, tube-type shock absorbers, stabilizer.

Dimensions
Length: 175.1"
Width: 69.6"
Height: 49.8" (coupe)

Instruments
Speedometer
Tachometer
Odometer
Trip odometer
Oil pressure
Water temperature
Fuel
Ammeter
Clock

Historical overview
- Four-wheel disc brakes were introduced in 1965 as standard equipment. There was a delete-disc option in 1965 only.
- 1965 was the last Corvette with fuel injection and the first with the "big block" motor. Introduction of the 396-cid engine (427 in 1966-67) made 1965 the first year since 1955 that engines of different displacements were available.
- Knock-off aluminum wheels were optional in 1965-66 (bolt-on style in 1967).
- 1967 was the last year for a closed coupe.
- Dual master cylinder brake systems were introduced to Corvettes in 1967.

Positive driver/owner factors
- Potent engines.
- Good pedal placements.
- Superb instrument locations.
- Superb instrument legibility.
- Telescopic steering column available.
- Simple manual top operation.
- Excellent ride quality.
- Air conditioning available.
- Ample headroom.
- Vent windows standard.
- Generous luggage space (coupe).
- Convenient handbrake location (1967).

Negative driver/owner factors
- No seat adjustment except fore-aft.
- Poor cabin ventilation (coupe).
- No dead pedal.
- Limited rear visibility (coupe).
- Difficult luggage access.
- Limited luggage space (convertible).
- Difficult battery access (some engines).
- Difficult spare tire access.
- Narrow footwells.
- Minor interior water leaks.
- Leaded fuel required.

1968-69

1969

Base price
1968: $4,320.00 (convertible)
1969: $4,437.95 (convertible)

Production
1968: 28,566
1969: 38,762

Brakes
Disc, front and rear

Transmission
Three-speed automatic; three-speed manual; four-speed manual.

Body style
Coupe with two removable roof panels; convertible with soft and/or removable hardtop.

Suspension
Front: Coil springs, tube-type shock absorbers, stabilizer, unequal length A-arms.
Rear: Transverse leaf spring, trailing arms, lateral struts, tube-type shock absorbers, stabilizer.

Dimensions
Length: 182.5"
Width: 69.0"
Height: 47.8" (coupe)

Instruments
Speedometer
Tachometer
Odometer
Trip odometer
Oil pressure
Water temperature
Fuel
Ammeter
Clock

Historical overview
• The 1968 Corvette featured a completely new exterior and interior. The wheelbase was the same as the 1963-67 models and the chassis components remained much the same.
• 1968-69 coupe models had twin removable roof panels; also, the rear window was removable.
• The Sting Ray designation was dropped for 1968, but revived for 1969 as Stingray.
• Wheel width was increased to seven inches in 1968, and to eight in 1969.
• 1968 was the last year for the 327-cid engine in Corvettes. The small block became 350-cid in 1969. Big block 427-cid engines were available both years.

Positive driver/owner factors
• Potent engines.
• Good pedal placements.
• Good speedometer/tachometer locations.
• Telescopic steering available (1968).
• Tilt-telescopic steering available (1969).
• Simple manual top operation.
• Air conditioning available.
• Convenient handbrake location.

Negative driver/owner factors
• No seat adjustment except fore-aft.
• Unsupportive seats.
• No dead pedal.
• Difficult battery access.
• Difficult spare tire access.
• Difficult luggage access.
• Limited luggage space.
• Narrow footwells.
• Limited shoulder room (especially early 1968).
• Reduced ride quality.
• Poor rear visibility (coupe).
• No vent windows.
• Leaded fuel required.
• Claustrophobic cabin.
• Poor secondary instrument location.

1970-73

1973

Base price
1970: $4,849.00 (convertible)
1971: $5,259.00 (convertible)
1972: $5,296.00 (convertible)
1973: $5,398.50 (convertible)

Production
1970: 17,316
1971: 21,801
1972: 27,004
1973: 30,464

Brakes
Disc, front and rear

Transmission
Three-speed automatic; four-speed manual.

Body style
Coupe with two removable roof panels; convertible with soft and/or removable hardtop.

Suspension
Front: Coil springs, tube-type shock absorbers, stabilizer, unequal length A-arms.
Rear: Transverse leaf spring, trailing arms, lateral struts, tube-type shock absorbers, stabilizer.

Dimensions
Length: 182.5" (184.6" in 1973)
Width: 69.0"
Height: 47.8" (coupe)

Instruments
Speedometer
Tachometer
Odometer
Trip odometer
Oil pressure
Water temperature
Fuel
Ammeter
Clock

Historical overview
- Compression ratios were dropped in 1971 to permit use of low-octane, low lead fuels.
- An alarm was first standard in 1972.
- 1972 was the last Corvette model with a removable rear window. Removal increased luggage space slightly in 1973 because a window storage tray wasn't needed.
- 1973 was the first Corvette model with a body-color front bumper. The chrome style was retained in 1973 in the rear.
- Steel beams for impact protection were added to Corvette doors starting in 1973.
- Big block engines increased from 427-cid to 454-cid for 1970-73 Corvettes.
- Starting in 1970, three-speed manual transmissions were no longer available.

Positive driver/owner factors
- Potent engines available (1970).
- Good pedal placements.
- Good speedometer/tachometer locations.
- Tilt-telescopic steering available.
- Simple manual top operation.
- Leaded fuel not required (1971-73).
- Convenient handbrake location.
- Air conditioning available.

Negative driver/owner factors
- No seat adjustment except fore-aft.
- Unsupportive seat design.
- No dead pedal.
- Difficult battery access.
- Difficult spare tire access.
- Difficult luggage access.
- Limited luggage space.
- Narrow footwells.
- Poor rear visibility (coupe)
- No vent windows.
- Leaded fuel required (1970).
- Claustrophobic cabin.
- Poor secondary instrument location.

1974-77

1975

Base price
1974: $5,765.50 (convertible)
1975: $6,550.10 (convertible)
1976: $7,604.85 (coupe)
1977: $8,647.65 (coupe)

Production
1974: 37,502
1975: 38,465
1976: 46,558
1977: 49,213

Brakes
Disc, front and rear

Transmission
Three-speed automatic; four-speed manual.

Body style
Coupe with two removable roof panels; convertible with soft and/or removable hardtop (1974-75).

Suspension
Front: Coil springs, tube-type shock absorbers, stabilizer, unequal length A-arms.
Rear: Transverse leaf spring, trailing arms, lateral struts, tube-type shock absorbers, stabilizer.

Dimensions
Length: 185.2" (185.5" in 1974)
Width: 69.0"
Height: 48.1" (coupe) (47.8" in 1974)

Instruments
Speedometer
Tachometer
Odometer
Trip odometer
Oil pressure
Water temperature
Fuel
Ammeter
Clock

Historical overview
- The first Corvette with body-colored bumpers front and rear was 1974.
- The rear bumper covers of 1974 Corvettes were two pieces. 1975-78 was one piece.
- Catalytic converters appeared first in 1975.
- 1974 was last year for big block engines.
- Convertible models were last available in 1975 (until 1986).
- High Energy Ignition system was first standard in 1975.
- Astro Ventilation was deleted after 1975.
- Sealed battery was first standard in 1976.
- Aluminum wheels were optional starting in 1976 (not available 1968-75).
- 1977 consoles allowed more radio choices.
- Cruise Control was first optional in 1977.
- Cloth seats were first standard in 1977.

Positive driver/owner factors
- Good pedal placements
- Good speedometer/tachometer locations.
- Tilt-telescopic steering available.
- Simple manual top operation (1974-75).
- Unleaded, low octanes fuels acceptable.
- Air conditioning available.
- Convenient handbrake location.

Negative driver/owner factors
- No seat adjustment except fore-aft.
- Unsupportive seat design.
- No dead pedal.
- Difficult battery access.
- Difficult spare tire access.
- Difficult luggage access.
- Limited luggage space.
- Narrow footwells.
- Poor rear visibility (coupe).
- No vent windows.
- Claustrophobic cabin.
- Poor secondary instrument location.

1978-79

1979

Base price
1978: $9,351.89
1979: $10,220.23

Production
1978: 46,776
1979: 53,807

Brakes
Disc, front and rear.

Transmission
Three-speed automatic; four-speed manual.

Body style
Coupe with two removable roof panels.

Suspension
Front: Coil springs, tube-type shock absorbers, stabilizer, unequal length A-arms.
Rear: Transverse leaf spring, trailing arms, lateral struts, tube-type shock absorbers, stabilizer.

Dimensions
Length: 185.2"
Width: 69.0"
Height: 48.0"

Instruments
Speedometer
Tachometer
Odometer
Trip odometer
Oil pressure
Water temperature
Fuel
Ammeter
Clock

Historical overview
• A new rear end design for 1978 featured a large rear window and increased luggage space.
• A limited edition Indianapolis pace car replica Corvette model was offered in 1978. Replicas had new style seats, special paint, and other unique features.
• 1978 was the Corvette's 25th anniversary and this was noted with special emblems.
• Wider 60-series tires were first available in 1978. They required trimming of the wheelwells at the factory for clearance.
• A dash glove box, the first since 1967, was added in 1978.
• Alarm systems were redesigned to include removable roof panels in 1978.
• The seat design introduced in the 1978 pace car replica was standard in all 1979s.
• AM-FM radios were first standard in 1979.
• Bolt-on front and rear spoilers, introduced on the 1978 pace car replica, became optional in 1979.
• Mirrored-glass roof panels, introduced on the pace car replica, were optional in late 1978 and all of 1979.

Positive driver/owner factors
• Ample luggage space.
• Good lateral seat support (1978 pace car replica, 1979).
• Good pedal placements.
• Good speedometer/tachometer locations.
• Tilt-telescopic steering available.
• Unleaded, low octane fuels acceptable.
• Air conditioning available.
• Convenient handbrake location.

Negative driver/owner factors
• No seat adjustments except fore-aft.
• No dead pedal.
• Difficult battery access.
• Difficult spare tire access.
• Difficult luggage access.
• Narrow footwells.
• No vent windows.
• Poor secondary instrument location.

1980-82

1982

Base price
1980: $13,140.24
1981: $16,258.52
1982: $18,290.07

Production
1980: 40,614
1981: 46,616
1982: 25,407

Brakes
Disc, front and rear

Transmission
Three-speed automatic; four-speed automatic (1982); four-speed manual (1980-81).

Body style
Coupe with two removable roof panels.

Suspension
Front: Coil springs, tube-type shock absorbers, stabilizer, unequal length A-arms.
Rear: Transverse leaf spring, trailing arms, lateral struts, tube-type shock absorbers, stabilizer.

Dimensions
Length: 185.3"
Width: 69.0"
Height: 48.0"

Instruments
Speedometer
Tachometer
Odometer
Trip odometer
Oil pressure
Water temperature
Fuel
Ammeter
Clock

Historical overview
- New front and rear caps with integral spoilers were added to 1980-82 models.
- Tilt-telescopic steering was first standard in 1980.
- Air conditioning was first standard in 1980.
- 1980 Corvettes sold in California had 305-cid engines instead of 350 because the larger engine wasn't certified for sale there.
- A plastic monoleaf rear spring was first used in 1981.
- A power seat base (driver side) was first optional in 1981.
- In 1981, Chevrolet opened a new Corvette assembly plant in Bowling Green, Kentucky.
- No 1982 Corvettes had manual transmissions.
- Cross-fire injection was first used in 1982.
- A Collector Edition 1982 model had a rear window hatch and other special features.

Positive driver/owner factors
- Ample luggage space.
- Good luggage access (1982 Collector Edition).
- Good lateral seat support.
- Good pedal placements.
- Good speedometer/tachometer locations.
- Tilt-telescopic steering standard.
- Unleaded, low octane fuels acceptable.
- Air conditioning standard.
- Convenient handbrake location.

Negative driver/owner factors
- No seat adjustments except fore-aft.
- No dead pedal.
- Difficult battery access.
- Difficult spare tire access.
- Difficult luggage access (except Collector Edition).
- Narrow footwells.
- No vent windows
- Poor secondary instrument location.

1984-85

1984

Base price
1984: $21,800.00
1985: $24,878.00

Production
1984: 51,547
1985: 39,729

Brakes
Disc, front and rear

Transmission
Four-speed automatic; four-speed manual with electric overdrives.

Body Style
Coupe with single removable roof panel.

Suspension
Front: Transverse monoleaf plastic spring, tube-type shock absorbers, stabilizer.
Rear: Transverse monoleaf plastic spring, five-link, upper and lower arms, lateral struts, tube-type shock absorbers, stabilizer.

Dimensions
Length: 176.5"
Width: 71.0"
Height: 46.7"

Instruments
Speedometer (graph and digital)
Tachometer (graph and digital)
Odometer
Trip odometer
Water temperature
Oil pressure
Oil temperature
Fuel
Volts
Clock
Instant and average fuel use

Historical overview
- There was no 1983 Corvette model.
- The 1984 Corvette body and chassis were completely new.
- The four-speed manual transmission introduced in 1984 had overdrives in the top three gears.
- The 1984 was designed with no external body seams.
- Electronic instrumentation was introduced to Corvettes in the 1984 model.
- The 1984 Corvette body design had a single, removable roof panel.
- A slightly modified Corvette with optional Z-51 suspension exceeded 1.0g in GM skid pad tests.
- Spring rates were reduced in 1985 due to complaints of ride harshness of 1984s.
- The cross-fire injection of 1984 was upgraded to port-fuel injection in 1985.

Positive driver/owner factors
- Exceptional handling.
- Exceptional sound system (Bose).
- Excellent underhood access.
- Ample luggage space.
- Good luggage access.
- Excellent seat design.
- Multiple seat adjustments.
- Good pedal placements.
- Comprehensive instrumentation.
- Tilt-telescopic steering standard.
- Air conditioning standard.
- Dead pedal standard.

Negative driver/owner factors
- Harsh ride (especially 1984 Z-51).
- Claustrophobic cabin.
- Difficult entry/exit.
- Difficult spare tire access.
- Very narrow footwells.
- No vent windows.
- Wear-prone interior surfaces.
- Poor speedometer/tachometer legibility.

1986-88

1986

Base price
1986: $27,502.00 (coupe)
1987: $28,474.00 (coupe)
1988:

Production
1986: 35,109
1987: -
1988: -

Brakes
Disc, front and rear

Transmission
Four-speed automatic; four-speed manual with electric overdrives.

Body style
Coupe with single removable roof panel; convertible with soft top.

Suspension
Front: Transverse monoleaf plastic spring, tube-type shock absorbers, stabilizer.
Rear: Transverse monoleaf plastic spring, five links, upper and lower arms, lateral struts, tube-type shock absorbers, stabilizer.

Dimensions
Length: 176.5"
Width: 71.0"
Height: 46.7" (coupe)

Instruments
Speedometer (graph and digital)
Tachometer (graph and digital)
Odometer
Trip odometer
Water temperature
Oil pressure
Oil temperature
Fuel
Volts
Clock
Instant and average fuel use

Historical overview
- Anti-lock brakes became standard in 1986.
- A convertible model Corvette was introduced in 1986. All 1986 convertibles were Indy pace car replicas.
- Aluminum cylinder heads were introduced in mid-1986.
- A sophisticated Vehicle Anti Theft System was standard in 1986-88.
- A high-mounted central stoplamp was added to 1986-88 Corvettes.
- Roller valve lifters were added to Corvette engines in 1987-88.
- A Callaway Twin Turbo engine package with 345-hp was available in 1987-88 through some Chevrolet dealers.

Positive driver/owner factors
- Anti-lock braking standard.
- Exceptional handling.
- Exceptional sound system (Bose).
- Excellent underhood access.
- Acceptable ride quality.
- Ample luggage space.
- Good luggage access.
- Excellent seat design.
- Multiple seat adjustments.
- Good pedal placements.
- Comprehensive instrumentation.
- Tilt-telescopic steering standard.
- Air conditioning standard.
- Dead pedal standard.

Negative driver/owner factors
- Claustrophobic cabin.
- Difficult entry/exit.
- Difficult spare tire access.
- Very narrow footwells.
- No vent windows.
- Wear-prone interior surfaces.
- Poor speedometer/tachometer legibility.

Appendix

Directory

This directory of addresses is by no means a complete list of all the companies that deal in Corvette-related activities or products. Rather, the companies mentioned in this book's text, plus some others which bear directly on subjects discussed, are listed. *Keepin' Track of Vettes* magazine prints an updated state-by-state nationwide Corvette parts and services directory of over 400 companies in one of its issues about every six months. *Vette Vues Magazine* is loaded with ads every month and it prints a list of its monthly advertisers in alphabetical order keyed to page numbers in each issue.

Driving schools

Skip Barber Racing School
Route 7
Canaan, CT 06018

Bob Bondurant School of High Performance Driving
Sears Point International Raceway
Highways 37 and 121
Sonoma, CA 95476

Jim Russell Racing Drivers' School
1023 Salinas/Monterey Highway
Salinas, CA 93908

Organizations

National Council of Corvette Clubs
P. O. Box 813
Adams Basin, NY 14410

National Corvette Restorers Society
6291 Day Road
Cincinnati, OH 45247

Stainless steel brake manufacturers

Dr. Vette
212 Seventh Street Southwest
New Philadelphia, OH 44663

Stainless Steel Brakes Corporation
11470 Main Road
Clarence, NY 14031

Vette Brakes
7490 30th Avenue North
St. Petersburg, FL 33710

Corvette product suppliers

American Custom Industries
5035 Alexis Road
Sylvania, OH 43560

Corvette America
P. O. Box 427
Boalsburg, PA 16827

Dr. Rebuild
P. O. Box 6263
Bridgeport, CT 06606

Distinctly Styled Corvettes
11408 Wheeling Drive
Tampa, FL 33625

Eckler's Corvette Parts
P. O. Box 5637
Titusville, FL 32780

Guldstrand Engineering
11924 West Jefferson Boulevard
Culver City, CA 90230

Mid America Corvette Supplies
1 Mid America Place
P. O. Box 1368
Effingham, IL 62401

Motor Sheep
5406 Complex Street #203
San Diego, CA 92123

Paragon Reproductions
8054 South Jennings Road
Swartz Creek, MI 48473

Zip Products
250 Commercial Centre
Mechanicsville, VA 23111

Racing organizations

Classic Sports Racing Group
P.O. Box 488
Los Altos, CA 94022

Historic Motor Sports Association
P.O. Box 30628
Santa Barbara, CA 93105

International Motorsports Association (IMSA)
P.O. Box 3465
Bridgeport, CT 06605

Sports Car Club of America (SCCA)
9033 East Easter Place
Englewood, CO 80112

Sportscar Vintage Racing Association
P.O. Box 2896
Lakeland, FL 33806

Vintage Auto Racing Association
5430 Townsend
Los Angeles, CA 90041

Vintage Motorcar Racing
1800 Market Street #20
San Francisco, CA 94102

Vintage Sports Car Club of America
116 Long Ridge Road
Bedford, NY 10506

Magazines

Corvette Fever
P. O. Box 44620
Ft. Washington, MD 20744

Corvette News
P. O. Box 7097
North End Station
Detroit, MI 48202

Keepin' Track of Vettes
1 Briar Lane
Spring Valley, NY 10977

Vette Vues Magazine
P. O. Box 76270
Sandy Springs, GA 30328

Victory Lane
2460 Park Boulevard
Palo Alto, CA 94306

Vintage Motorsport
P. O. Box 2955
Lakeland, FL 33805

Recommended books

The sheer number of books written about Corvettes attests to the marque's popularity. Here's a very brief listing of very good examples.

Adams, Noland. *The Complete Corvette Restoration & Technical Guide—Vol. 1, 1953 Through 1962.* Princeton, New Jersey: Princeton Publishing, Inc., 1980.

— — — *The Complete Corvette Restoration & Technical Guide—Vol. 2, 1963 Through 1967.* Princeton, New Jersey: Princeton Publishing, Inc., 1987.

Antonick, M.B. *The Corvette Black Book.* Powell, Ohio: Michael Bruce Associates, Inc., 1978, 1980, 1984, 1985, 1986, 1987.

— — —*Illustrated Corvette Buyer's Guide.* Osceola, WI: Motorbooks International, 1986.

Lamm, Michael. *The Newest Corvette: From A Through Z-52.* 2nd ed. Tucson, Arizona: HP Books, Inc., 1986.

Ludvigsen, Karl. *Corvette: America's Star-Spangled Sports Car: The Complete History.* 2nd ed. Princeton, New Jersey: Princeton Publishing, Inc., 1978.